D0061554

Our Galactic Visitors

THE EXTRATERRESTRIAL INFLUENCE

The Sacred Science Chronicles: Volume V

ROBERT SIBLERUD

New Science Publications
9435 Olsen Court
Wellington, CO 80549
(970) 568-7323

OUR GALACTIC VISITORS:
The Extraterrestrial Influence
Copyright 2008 by Robert Siblerud
All Rights Reserved

Published by:
New Science Publications
9435 Olsen Court
Wellington, Colorado 80549

Printed in Canada

ISBN: 978-0-9666856-6-4
Library of Congress Catalog Card Number: 2001012345

Acknowledgements:

A True Pioneer: **Leo Sprinkle, Ph.D.**
Editing: **Margaret Shaw**
Copy Editing: **Shirley Parrish**
Reference Material: **Franklin and Carolyn Carter**
Cover Art: **Ron Russell**
Layout: **Pat Alles**

TABLE OF CONTENTS

FOREWORD

R. Leo Sprinkle, Ph.D.

Professor Emeritus
Counseling Services
University of Wyoming

Welcome, dear reader, to a treasure of data, information, knowledge, and wisdom regarding extraterrestrials. Dr. Robert Siblerud has gathered a mountain of evidence about our galactic visitors and their influence on human development. That in itself is a monumental task. He has organized the evidence in a rational format and presented it in a readable style that whets the intellectual appetite of any seeker of truth.

Dr. Siblerud has been able to describe and interpret physical and astronomical data; biological and DNA information; psychological and cultural knowledge; spiritual and channeled wisdom that has been given to us by our galactic friends. This is not unusual for the author, as he has written four other books on "sacred science," topics including ancient civilizations, mystical societies, the science of the soul, and the unknown life of Jesus. He has demonstrated, in these earlier volumes, his ability to organize and offer a coherent, comprehensive, but compact interpretation of modern texts and ancient documents.

A cautionary comment to any seeker of truth: Dr. Siblerud may be pleasant and mild mannered in his social behaviors, but he is bold in his willingness to go beyond contemporary boundaries of religion, science, and politics. The fears of fervent fundamentalists may be barriers for some readers who are puzzled about the agenda and activities of our galactic visitors.

Thus, the reader who experiences doubts about these writings may wish to check and compare the sources of literature. Or the reader may wish to use the "applied kinesiology" method in these writings (a la D. R. Hawkins, M.D., Ph.D, *Power vs. Force*). Or the reader may wish to adopt a "wait and see" attitude regarding the transformation of humanity (a la Owen Waters, *The Shift: The Revolution in Human Consciousness*).

For example, if a reader distrusts or doubts channeled communications, then he or she may wish to consult the book *Channeling* written by college professor John Klimo, PH.D. Dr. Klimo has shown that everyone engages in "open channeling" (understanding through intuitive awareness). The writings of Lynne McTaggart (*The Field: The Intention Experiment*) demonstrate that the holographic universe allows for instant communication between all points of the universe.

The question for any doubtful reader is not whether ESP or intentional influence exists, but the questions is: at what level of "truth," and how much "truth" am I able to apprehend within my conscious awareness? The Map of Consciousness (Hawkins, *Power vs. Force*) indicates that there is some "truth" at every level of consciousness. The question for humanity is whether we are ready to minimize "duality" and maximize "unity" in our search for further truth about ET presence and our current potential participation in cosmic cultures?

Dr. Siblerud provides a possible answer to the controversy between creationists and evolutionists. Some theologians have assumed a dogmatic position: divine creation of humans is the only explanation for the origin of humans. Some scientists have assumed a dogmatic position: biological evolution is the only explanation for the origins of humans. The thesis/antithesis model of ideas, developed by Immanuel Kant provides a method of resolution: neither the creationists nor the evolutionists have the complete answer.

Researchers (e.g. Marshall Klarfeld, Lloyd Pye, Zecharia Sitchin) have shown that the available evidence indicates that "Adam" (early humans) have both the genes of primates (homo erectus) and star seeds (ET genetic code). It seems that we humans were created by a female geneticist, Ninhursag, a scientist of the Anunnaki, a race of gods who established the Sumerian civilization. Now the extreme patriarchs of both science and religion can come together to commiserate: the Goddess/God of creation has many mansions and many families.

This volume offers a challenge to the reader. If humanity has originated and evolved through the influence of many extraterrestrial groups, and if humanity is facing a crisis in physical and spiritual changes, including the current "cover up" of the ET presence, how should we respond? Dr. Siblerud concludes in the book, "We are approaching the end of a great cycle and we have the opportunity to evolve to a higher dimension." May the efforts and awareness of the author and awareness of the readers be merged, and may we all be of service to others as well as to our Higher Selves. May we all experience the transition of human consciousness.

In Love and Light,

Leo Sprinkle, Ph.D.

PREFACE

When one understands the influence on humanity that extraterrestrials have played, it is unbelievable how well the secret has been kept to maintain our ignorance. As you will discover in this book, galactic visitors have influenced our human family in almost every aspect of life, including religion, science, wars, politics, health, and genetics. We are the offspring of our galactic visitors and carry their genetic lineage. Humanity is beginning to awaken to the truth of extraterrestrials. Most galactic civilizations want us to know the truth about them and our heritage but realize Earth humans are evolving. Many of us are not ready for this profound knowledge.

Throughout this book, mention will be made of several galactic civilizations that are trying to retard humanity's evolution. The reasons for this will be explained. These are the groups working with the Secret Government to keep us in the dark and hindering human evolution. There is a force of unenlightened beings, both ET and humans, that is trying to control us and the planet. If ETs exist, many rightfully ask, why don't they just land on the White House lawn and announce themselves? As one delves into the ET connection, it will be discovered that our government has shot down a number of ET craft. The extraterrestrials are very reluctant to announce themselves because of our warlike consciousness. It would provide an additional excuse to build up arms against the perceived threat. The enlightend extraterrestrials have an agreement of noninterference through their organization called the Galactic Federation. However, it is the unenlightened ETs that are influencing our Secret Government to keep our evolution retarded.

Most of our galactic visitors tell us that we are entering a window of time that will allow us a quantum leap in evolution. This window corresponds to the Mayan Calendar year of 2012, the year we enter a new era. The window is small and we only have this rare opportunity once every 26,000 years. They tell us we have the chance to evolve from the Third Dimension into the Fourth and Fifth Dimensions. On a subtle level our galactic friends are doing what they can to raise our consciousness to allow us the opportunity for this evolution. At one time, all these advanced galactic civilizations have been at our level of evolution as well. Only when they began to understand the Law of One were they able to evolve into these higher dimensions. The Law of One is unity consciousness, meaning that we are all one and connected to each other and everything. When we understand this principal, peace will come and we will celebrate our diversity. Quantum physics has provided the science underlying this law showing that we are all connected. We are rapidly approaching this window of enlightenment, and as many have noticed, there has been a rapid increase in consciousness regarding environment, health, and even peace, despite the Iraq war, which may be contributing to the raised consciousness about peaceful alternatives. Our galactic friends are confident that we can join them in this new dimension and are pulling for us to make this leap in consciousness.

I do not pretend to be an expert in UFOs and extraterrestrials. It is from the experts that I have gathered knowledge for this book. My *modus operandi* has been to gather information from the experts and synthesize it into book form. This is the fifth book of the Sacred Science Chronicles. My introduction to UFOs and extraterrestrials began in the 1980s when I attended the Rocky Mountain UFO conferences at the University of Wyoming. The conferences were convened by psychologist Leo Sprinkle, Ph.D. who was a professor at the University of Wyoming. Dr. Sprinkle conducted a gathering to assist people who had been abducted by extraterrestrials so they could better deal with the trauma of their abduction. He discovered that most abductees had a tremendous spiritual growth following their experience. Leo always had a positive attitude about abductions that really helped people cope with their trauma.

While I was director of the International Association for New Science (IANS), we conducted many conferences in new science areas including UFOs. One year we gathered the top UFO researchers for a retreat at St. Malo's Lodge in the shadows of Colorado's 14,000 foot Mount Meeker, followed by a conference in Denver. In the mid-1990s, IANS took over Dr. Sprinkle's conference at the University of Wyoming. Our core group evolved into the Institute for UFO Research headed by Franklin and Carolyn Carter, who continued to put on the conferences until the early part of this century. Personally, I was always more interested in the extraterrestrials than the mechanics of UFOs. For many years I have wanted to bring together as much knowledge as I could regarding our galactic visitors and the fruition of that desire is this book.

I don't expect the reader to believe everything in the book. Often, I will present opposing views and let the reader discern his or her own truth. I have gathered the facts to the best of my ability and hopefully have presented them accurately. Welcome to the greatest story never before told.

Robert Siblerud

Chapter One

SEEDING THE GALAXY
Meeting Your Ancestors

Where did we come from? This question has been debated by theology and science for ages. Was man created on the sixth day of creation as Genesis states? Science has a difficult time with that verse, as it does with much of the Bible. Science claims humans evolved from the primates, but there are missing links to that hypothesis. Where did the primates come from, we might ask? Science and theology believe they have most of the answers regarding man's appearance on the planet Earth. As we shall see in this book, both science and theology are partially right and partially wrong. We will discover that man was created by a small "g" god in nine months, not six days. We will also discover that man did evolve from the primates, but with the help of a missing link, a small "g" god. The gods of mythology and the Bible are one and the same. They are extraterrestrials and they are our creators.

Humans are on a spiritual path of evolution. Our souls are evolving, and by understanding our extraterrestrial ancestors' role in our progress, we can begin to understand the many mysteries surrounding our creation and the mysteries of our evolution. This book is not going to try to prove the existence of UFOs and extraterrestrials. Any scientist worth his or her salt, after studying the evidence surrounding this phenomenon, can come only to the conclusion that UFOs and extraterrestrials do exist. We humans are about to enter an extraordinary phase in our evolution, joining our galactic family. As we shall see, our window of opportunity is rapidly approaching, and hopefully this book will provide you with insight on how this is going to happen.

1

One can rightfully ask, if traditional science does not acknowledge the existence of intelligence beyond our planet Earth, where does the information come from to enable one to write a book such as this? Traditional science has been trapped in a cage, and those scientists who try to escape are refused research funding and in all likelihood will lose their jobs if they look for answers outside their academic cage. As we shall see, there are powers that want to suppress this new science of ufology and keep the lid on any other new paradigm of science. There have been many pioneer scientists who have broken from the confines of the scientific cage and have done excellent research regarding UFOs, extraterrestrials, the spiritual world, and the paranormal. It is from these sources that evidence is available to enable one to write books such as this. If one looks at the Bible, ancient myths, and ancient writings, some of the answers lie in this evidence. A number of people have been in direct contact with extraterrestrials, including our U.S. government. Today, it is the Secret Government that is maintaining communications with the extraterrestrials.

Most extraterrestrial civilizations want the general Earth population to have information about this forbidden knowledge and through various means have tried to give it to us. However, enlightened or positive extraterrestrials are bound by a code of non-interference which prohibits them from interfering directly with human activities and their evolution. They do have indirect methods that they use to share knowledge. On the other hand, there are several renegade or negative extraterrestrial groups that have interfered directly with humanity by way of abductions and providing highly-advanced technology, mainly to the Secret Government. These negative extraterrestrials have hidden agendas which are not in the best interest for humanity's evolvement. As we shall see, the Secret Government has used some of this knowledge to control worldly events, hindering our evolution. Because of their power, the negative extraterrestrials have been able to manipulate their own agenda. We will explore both the positive and negative extraterrestrials and their influence on humanity. The positive extraterrestrial civilizations are trying to warn Earth humans that time is running out to enter this window of evolution to the Fourth and Fifth Dimension, which is why it is so important to understand the dynamics of extraterrestrial involvement that is affecting our lives today.

EARLY GALACTIC CIVILIZATIONS

We cannot go to the scientific journals to research early galactic civilizations. However, a handful of people have been able to communicate directly and indirectly with extraterrestrials to acquire the history of these civilizations. Communication between extraterrestrials is done with the mind, something like mental telepathy. Some extraterrestrials have special devices attached to the head that allow them to understand foreign languages of various races. Their advanced genetic makeup also allows them the gift of communication through their minds, their primary mode of communication. Some advanced earthlings also have the correct DNA that allows them to communicate with extraterrestrials.

One of these individuals is Anna Hayes*, an American extraterrestrial liaison, who has written a series of books entitled *The Voyagers*. Anna says the information she receives is received through a process called "Keylonta," a language of light, sound, symbols, and energy, which is not a channeling process. It is considered the language of the stars. Through this technique, Anna claims to have the ability to communicate with extraterrestrials who have downloaded remarkable knowledge to her. The information is like a graduate course on extraterrestrials and evolution that needs to be shared with all of humanity. Another source of history about early galactic civilizations comes from Lyssa Royal in her book *The Prism of Lyra,* describing the birth of the humanoid race that dovetails nicely with the Hayes material. Much of Lyssa's material is used in this chapter. She claims to have received the information from an extraterrestrial source through mental telepathy or channeling. It is quite surprising how consistent her information is when compared to other sources. Besides these two sources, a number of other sources indicate that there are other extraterrestrial civilizations which have played a major role in the lives of humanity and the planet Earth. This will be discussed later in the book.

* Anna Hayes is now writing under the name of Ashayana Deane

Lyra

Most extraterrestrial sources concur that the birth of the humanoid race was in the star system Lyra. All humanoid races in our galaxy family are genetically connected to Lyra. The creators of galactic humanity on Lyra were etheric in nature, tall with two arms, two legs, a head, torso, and large eyes. Both the Hayes* and Royal sources called these creators the Founders. As the Lyrans fragmented their consciousness, it began to solidify into matter, creating a Third Dimension physical reality and a prototype physical race into which the majority of humanoid consciousness would incarnate.

An organizational code (DNA) created a consistent humanoid carbon-based body that became a vehicle for humanoid consciousness. The Founders used a naturally occurring code that assisted in creating versions of themselves in both the physical and non-physical states. The created forms symbolically reflected the aspects of a polarized universe. The Lyran creators wanted to manifest different dimensional aspects of themselves and to produce a root form that would bring diversity into their new reality. As this new humanoid form diversified, the Founders became aware that through evolution all life would return to the higher dimensional creators and then back to the Source. As fragmented as this consciousness was, sometimes beyond recognition, it would eventually rejoin the Source at all levels – physically, emotionally, mentally, and spiritually.

As these new races were developed, planets were chosen within the Lyran star system for these created races to inhabit, according to Lyssa Royal's source. These planets began to develop primate life, and the creators seeded these developing primates with plasmic energy within their DNA. As evolution progressed, the primate/humanoid possessed genes to sustain a higher vibration with a Three Dimensional consciousness. These incarnations progressed on several planets and each consciousness became attached to a specific planet with a template of triad qualities – positive, negative, and integration of the positive with negative.

As the races evolved, they achieved technology for space travel and began to mix with other cultures. The Lyran races developed an expanded philosophy, a strong social development, and advanced technology. As they interacted with other cultures, the polarities began to solidify and their own polarities began to integrate. The feminine became more masculine and the masculine more feminine. As they continued their development, the new races lost contact with their Lyran creators, and the new creation took on lives of their own.

The first Lyran group to develop a non-Lyran species was the Vegan civilization, which developed a very different philosophy and spiritual orientation from the other Lyran races. They were initially a negative-orientated civilization originating from the negative pole of Lyra, and they adopted a philosophy of service to self, or contraction. Other civilizations that developed out of Lyra were positively orientated and expansive. Vega is the brightest star (Alpha) in the Lyra constellation.

Friction grew between the Vegan and Lyran races because of the two polarities. Neither group was right nor wrong, because they simply viewed the same idea from a different point of view. This polarization began to grow exponentially.

Another civilization emerged from the Lyran race, which was the apex of a symbolic triangle of integration. They played out the polarities of both positive and negative and became known as the Apex planet during their development. Their civilization acquired traits from both the Lyran and Vegan polarities. Genetically, they were a mixture of the two races.

Diversity became widespread on Apex, and the inhabitants did not coexist peacefully. Because of this separation, they became fixated on their polarity. As a result of this turmoil, they allowed pollution and nuclear weapons to nearly destroy their planet. A nuclear war did break out that nearly destroyed the entire planet except for the survivors who went underground. The survivors underground went through a great transformation and were propelled into an alternative dimension. The ancestors of these survivors, called the Grays, are playing a role on Earth today.

The other two civilizations, Lyra and Vega, continued to evolve. They began to colonize other star systems, with Vega founding colonies on Altau and Centauri. As galactic humanity spread, it carried the seeds of experience and the challenge of polarity. Integration of this polarity was still the main goal. The dynamics between the three civilizations became unpredictable, but the Founders knew that divine order would eventually come out of chaos.

Sirius

Conflicts became more frequent between the Lyrans and Vegans. Knowing the challenge they had with their polarities, both civilizations sent colonizers to the star system Sirius to integrate their polarities. The non-physical Sirius had evolved into a dimensional density that was able to support the life forms from both Lyra and Vega. Early Sirius was suitable to create realms where all manifestations of consciousness could exist and was known as the Eden of Sirius. Sirius became an important symbol for the galactic family because of its trinary star system that reflected the goal of integration of the positive and negative polarity. Sirius was to become a place for some of the earliest genetic engineering in the galaxy, following the path of the Founders.

The Vegans who came to Sirius were very masculine, and their philosophy was domination to control their evolution. Because of this strong negative polarity, it became difficult to maintain a connection with the Fourth Dimension. Once the primate-like species DNA became compatible with the Vegans on Sirius, the Vegans began incarnating. Almost immediately, the new Sirians lost their memory connection to Vega and remembered nothing of their origin. They created a society ruled by their desire for domination over others and the universe around them.

As the Vegans were developing on Sirius, a group of Lyran colonists came to the planet. The Lyran newcomers had a positive polarization toward service to others and were very interested in healing. The combination of positive and negative energy of these two systems created an environment of tension among the genetic heirs of these two opposite races. Finally, the Elders of Sirius had to intervene, and it was decided to colonize another star system, Orion, which had the correct electromagnetic properties for balancing their civilization. The negative Sirians were in a state of denial regarding the conflict caused by their negativity and were still in a state of forgetfulness. On Orion, the positive Sirians were now able to directly influence the negative polarity. Other positives from Lyra came to Orion and galactic history was born.

Because the positive Sirians had a passion for physical healing, they had aligned themselves with extraterrestrials from Arcturus who were orientated toward emotional healing. Out if this alliance arose the Sirius/Arcturus matrix that carried a holistic energy of healing mind, body, and spirit to all the planets in the galaxy. Earth has benefitted greatly from the Sirius/Arcturus matrix. A small percentage of positive Sirians decided to incarnate in the Third Dimension in cetacean form, giving rise to the dolphins and whales. The purpose of this incarnation was to transmute energy of physical polarity to help bring harmony to the planet.

Extraterrestrials around the universe claim that Sirian energy is the most widely used upon the Earth when compared to all other galactic energy. Sirius is the brightest star seen from Earth and is the second nearest. Energy from Sirius was used by many cultures, especially the Egyptians during the early Egyptian dynasties. The Sirians had chosen a frequency density that was visible to Third Dimensional humans. The early Egyptian gods of Isis, Osiris, and Anubis were gods from Sirius, who bestowed knowledge of astronomy, medicine and agriculture. The Mayans also had a close relationship with the Sirians. Many Mayans were Sirians who desired to experience physicality and so they incarnated upon Earth. The Sirians shared with the Mayans knowledge of technology that would transmit matter into pure energy or consciousness. Following their Earth mission, the Mayans vanished by transmuting physical reality. The Sirians have been a guiding force for development of civilization upon Earth. However, the negative Sirians also had their earthly influence. They were instrumental in promoting misuse of the knowledge given to humanity, and some Egyptian priesthoods worshipped the negative force in their temples.

Civilizations on Earth have been influenced by the negative Sirians (and negative Orions) through the secret organization of the Illuminati, which has great influence on the Secret

Government. These negative Sirians are identified by the need to control, and they have interfered with Earth's development from the start. The Illuminati Sirians are only a small percentage of Sirian consciousness, and they are terrified of nonexistence. Negative UFO experiences, cattle mutilations, and Men in Black are associated with the negative Sirians. Often they generate more fear than actual damage. Because of their fear of nonexistence, the negative Sirians have been prevented from transitioning into the Fourth Dimension. By keeping humanity in fear, Earth's inhabitants will also be prevented from making this transition into the Fourth Dimension.

Apart from this negative interference, the early Sirians were quite talented at genetic engineering. These early Sirians placed a latent DNA code within humans that will be triggered when human consciousness reaches a certain frequency or vibration. Earth's destiny is to inherit the Sirian triad, the integration of polarity.

Orion

As the civilization of Lyra tried to achieve integration within the Vegan and Sirius star systems, the conflict of polarities that began to play out spilled over to the Orion star system and eventually evolved into a race war. The negative Orions continued to serve themselves, and by doing so, they rationalized that the whole was being served. Therefore, they needed to dominate. Genetic manipulation of blood lines was used to concentrate or dilute power, depending on their need. Fear became a dominant characteristic of the negative Orions, especially of those who were different. On the opposite end, the positive Orions were so entrenched in the concept of serving others that they became subservient victims, even at the expense of self.

Technological advancement was the primary gift of the Orion civilization, even when they were in a state of spiritual conflict, creating a dangerous situation in the hands of the spiritually unevolved. Realizing this predicament, many souls would incarnate into the positive civilization and then into the negative civilization, in the hopes of integrating their individual polarity.

The Orion Empire had devised methods to control the astral body, so death was no longer a freedom. Few souls found their way out of Orion. However, a small percentage of souls were able to leave their bodies following death and incarnate into the Earth reincarnation cycle. These souls became lost in Earth's mass consciousness, losing their own identities, but at least free from the holds of Orion. After entering the Earth's reincarnation cycle, they continued to play out the Orion drama unconsciously within their soul patterns. If the negative Orions reincarnated in the Earth cycle, these souls would bring along their desire for control.

After many incarnations on Orion, many souls embodied the dreams of most Orion races to have integration without fear and hatred. They began to learn that one must integrate the positive with the negative, so that both sides must release fear in exchange for love. Free will choice was an important attribute for the inhabitants of Orion. To preserve their civilization, scientists had engineered a latent code into their genes that would trigger preservation of the society when they were about to self-destruct. They had chosen the planet Earth to help transmute the polarized Orion energy.

Throughout Earth's history, the Orion drama has been played out in an attempt to balance polarity. Memory patterns of Orion were found in the fall of Atlantis, the Roman Empire, and ongoing religious wars, replaying the attributes of victim, domination, and resistance. Earth today still sees these dramas unfold. Some Orions see Earth as a threat and want to control our planet, keeping humanity unempowered. However, Earth is about to go through a transmutation and integration process that will affect the entire galactic family. As Earth awakens, humanity will have a chance to claim responsibility for both self and the whole.

Pleiadians

Even during the early era of Lyran development, friction between the two polarities manifested. The feminine polarity involved the concepts of intuition and allowing, and believed the path to integration was through inner development. The masculine polarity wanted to dominate, thinking that was the path to evolvement. Dissension between the two polarities grew. A group of Lyrans wanted to develop their culture far away from the negative influence of Lyra. They discovered Earth, which was abundant in natural resources. After several generations, they realized they could not adapt to the Earth's physical environment and electromagnetic energy. During their stay on Earth, they began to experiment with genetic material from primates to help them adapt, and some primates were injected with Lyran DNA. Conflicts began to rise between the Earth-Lyrans and later arriving negative Lyrans. To avoid these conflicts, the Earth-Lyrans discovered a new home in the Pleiades and colonized that star system. Their hope was to create a civilization based on harmony, truth, and unconditional love. The early Pleiadians (Earth-Pleiadians) possessed great intuition and a desire to create a community lifestyle where the whole was more important than the self. After generations of development, the community-orientated Pleiadians favored peace and tranquility so much that negativity became invalidated. As a result, there was no conflict nor learning from conflict resolution, and a great void developed. They had become isolated in the universe. They consulted their Lyran forefathers about their dilemma and were told about the anguish ocurring on Orion.

Being of a service consciousness, the Pleiadians developed a plan for assisting Orion in their struggle that, at the same time, allowed them to evolve spiritually for their own benefit. Some Pleiadians incarnated into Orion, but found themselves entrapped in the Orion struggle. Once in the reincarnation cycle, it was almost impossible to escape it. Others continued to incarnate within the Pleiadian system and tried to help by limiting the expansion of the Orion Empire.

The Pleiadian's fight against negativity continued. However, instead of finding the truth within, they perpetuated the hatred of negativity. Only after Orion destroyed one of the Pleadians' populated planets did they decide to withdraw from the Orion struggle. Needing another mission with higher purpose, they decided to return to Earth.

Under direction of the Founders, the Lyrans started the

Inception Project, assisted by the Sirians. The project required genetic material from Earth plus extraterrestrial genes, so they contacted the Pleiadians because they had both. At first, the Pleiadians were reticent about becoming involved with Earth because of their previous experience. Because the Pleiadians had incorporated primate genetics into themselves, the Lyrans wanted Pleiadian DNA to develop terrestrial species on Earth, giving the Pleiadians a chance to come to terms with their views on negativity.

The plan was to transfer Pleiadian DNA into Earth species and create a humanoid race. The Pleiadians would be the closest ancestors of Earth humans, and they would be allowed to assist in the development of Earth's humanity. Their role was to observe humanity's development and to interact occasionally to keep them on the evolutionary track, providing the Pleiadians a chance to learn about human negativity. The Pleiadians agreed to the Inception Project, resulting in an interaction with nearly every primitive culture on Earth. Drawings of space beings and space-craft have been found on many cave walls around the world, and ancient writings document the gods who came from the sky.

Humans began to give up their personal power to these godlike entities, and some of the Pleiadians enjoyed this attention and resulting power. Soon they began to manipulate humans and created an abnormal desire for attention. Many ancient myths describe the jealous gods. Other extraterrestrial groups began to visit Earth, creating great resentment among the Pleaidians. Seeing that they were developing negativity, the Pleiadians knew they had to do something about it. They pulled back. Realizing that Earth humans were no longer children, the Pleiadians allowed humanity to make their own choices.

As Earth humans developed, the Pleiadians have watched very closely for the critical mass to activate the DNA code for preserving the species. Since the 1940s, the beginning of the nuclear age, both physical and non-physical extraterrestrials have been monitoring humanity and communicating knowledge to humans in subtle ways. At first, they warned us of great natural cataclysms if humans did not change their ways. During the early 1980s, there had been a great shift in mass consciousness for increasing responsibility toward the Earth and humanity. The Pleiadians realized they had created a karmic cycle for themselves by interfering with Earth's development. They needed to break the pattern of interference, which was the most fearsome goal facing them.

The Pleiadians are the most similar in appearance to Earth humans in Fourth Dimension form, and they can walk among us without a disguise. Many people on the planet Earth feel a strong connection to the Pleiades, and perhaps the Pleiadian historical connection explains this connection.

Arcturians

The most evolved civilization interacting with Earth is from Arcturus, symbolizing our future selves and future society. They are here to serve humanity. Arcturians are the example of integration of the positive and negative. Humanity is evolving into nonphysicality, and the goal is to achieve a consciousness similar to the Arcturian mass consciousness. The Arcturians consider themselves a group matrix committed to the goal of consciousness evolution.

According to Lyssa Royal's source, because Arcturians are Sixth Dimensional and were never Third Dimensional, they have chosen to learn about physicality through human beings. On the other hand, according to author Norma Milanovich's Arcturian source that will be discussed in another chapter, the Arcturians are Fifth Dimensional and had evolved to that dimension from the Third Dimension. Often they have manifested to humans as angels. They also have the ability to manifest themselves according to the belief system of the individual with whom they are interacting. For the more traditional, they will appear as angels. The Arcturians also interact with Earth's unseen kingdoms whose evolutionary path differs from humans. They are the higher aspect of the deva kingdom.

A few Arcturians have chosen to serve humanity by experiencing physicality. They have "walked in" an already-existing body to help humanity's evolution. They have no karmic obligation. Souls of humans who are in great emotional pain will enter the Arcturian realm for healing after death. It is said that souls will go to this realm for healing to begin their afterlife.

Arcturus is the first brightest star located in line with the handle of the Big Dipper. It has become more of a realm than an actual Three-Dimensional place. In the Arcturian realm, each soul who has had a traumatic death or unbearable life is healed and rejuvenated in Arcturus. Fortunately for Earth, we are connected dimensionally with Arcturus, and all who incarnate on Earth must pass through the Arcturian realm. They also heal those souls who are about to be born. Often, birth is more traumatic than death. They prepare the nonphysical consciousness for the intense life it may encounter on the physical plane. At death, the human consciousness passes through the Arcturian realm and is cared for and nurtured until it is ready to experience its next reality.

Arcturus is in the Sixth Dimension, often thought of as the Christ or Buddha vibration, according to Lyssa Royal, while the Norma Milanovich source says that in the Arcturian Fifth Dimension they are in constant contact with Christ. The light at the end of a tunnel in a death experience is an Arcturian vibration. The light is the symbol of the Higher Self, where one merges with Higher Self for healing. The Arcturian energy is a frequency of healing, creation, and evolution, and has been with Earth since the beginning. The Arcturian vibration is like the glue that holds it all together. Arcturus and Sirius formed a partnership for healing, where Sirius facilitates physical healing, and Arcturus' energy works with emotional healing. Arcturus represents the integration of polarity, whereas Orion is the separation of polarity, representing conflict and humanity's polarity. In the early development of Earth, the Arcturians densified themselves to be perceived temporarily by the Lemurians who were taught the skills of healing. Statutes on Easter Island pay tribute to the Arcturians. Through the loving efforts of the Arcturians, humanity is evolving from the Third Dimensional world to the Fourth and Fifth Dimension.

Zeta Reticuli

The Zetas are the most recognized extraterrestrial group on Earth, commonly called the Grays. They have an interesting history that helps explain their earthly presence. Their roots begin in the Lyran star system with the inception of the Apex planet under the guidance of the Founders. On the Apex planet, polarity was expressed through individualization that led to separation from the whole. The evolution of technology progressed rapidly, more rapidly than their spiritual evolvement, which prevented a peaceful coexistence of its inhabitants. Because of this separation, wars broke out, and the planet Apex became very polluted. Eventually a nuclear war broke out which made the planet surface uninhabitable. Survivors on Apex saved themselves by going underground where they developed an underground society. To survive as a civilization, they needed to integrate so that wars and pollution would not destroy the remaining culture. They decided to take a drastic step and force integration through restructuring of their reality. They became so intellectually developed that their craniums were too large to pass through the birth canal. Genetic engineering was undertaken to replace the birthing process, which was a lifesaver, because they had found themselves sterile after the surface was destroyed.

The Apexians knew they did not want a civilization like the one they had, so they rigorously controlled the genetics of the future society. Their first priority was to genetically alter emotional expression because their desire was order, not passion and chaos. The brains were altered so there was only a consistent chemical output to an external stimuli. This allowed them to detach from their egos. As a result, the Apexians developed a group mind with the loss of all individualization. Because of planetary radiation and cloning, everybody began to look alike with little physical variance. In order to adapt to the underground, their bodies became smaller through genetic engineering. Since there was no ultraviolet light underground, their eyes began to respond to different frequencies of the visual spectrum, and their pupils mutated to cover the entire eye. The eyes became enlarged to allow a greater area to gather light. Fresh foods became a problem, so they adapted their bodies to absorb a certain frequency of light for nourishment. Salvaged plants and underground minerals also helped with their nourishment. Because of these unusual dietary patterns, their internal organs began to atrophy.

As a result of the nuclear war, the blast folded the space surrounding Apex, and it emerged on the other side of a dimensional doorway. The Apexians were underground for thousands of years and had not realized the planet had shifted its position in relation to space and time. Once they emerged, they began to master the science of folding time and space. Because of the dimensional shift, they were now located in the vicinity of Zeta Reticuli 1 and Zeta Reticuli 2 in the Reticulum Rhomboidasis star group.

The Apexians had reconnected with their creator and continue today to carry out their evolution and help with galactic evolution to honor the wishes of the creator. Because of this transformation following the destruction on their planet, they became "One people, reflecting the whole." Today, many refer to them as the Zetas.

Because of all of their genetic engineering, the Zetas have become an endangered species and have a long way to go to strengthen their genetic line in order to save their race. They have become severely inbred and stagnant in their spiritual growth. Even though the race is dying, the oversoul wishes to incarnate into the Third Dimension. As a result, they have deliberately kept themselves from transitioning to the Fifth Dimension in order to leave a seed of themselves that can continually reproduce. If they are able to, it will aid the galactic whole in evolution.

The Founders had told the Zetas that Earth possessed a gene pool from many types of species going back to the inception of the Lyran race. In the 1940s, when humans began to gain technology that could destroy Earth, the Zetas became quite attracted to Earth. They had the ability to time travel and gather genetic material from any time period of Earth. However, they needed the genetic material from this period of Earth's history when civilization was on the brink of self-destruction and transformation. This DNA will assist in their goal of integration. By interacting with Earth, the Zetas can heal their past and change their future.

This is the background underlying Zeta abductions in which they carry out genetic experiments. Most abductees consider themselves victims, and thousands have been terrorized when taken from their beds. In another chapter we will discuss soul agreements for these abductions. Often the abductee will have a spiritual transformation following an abduction. The Zetas are looking for human characteristics that they bred out of their race ages ago. They want to recapture emotions and, therefore, seek people who react in a variety of ways to external stimuli. To monitor neurochemical secretions, they implant organic probes into the abductee's head, nose, eye, or ear cavity. The probes catalog neurochemical data and are removed periodically for study. Following death, the remaining probe is naturally absorbed by the body.

The Zetas have not parented children for eons. This is why women are abducted and asked to hold hybrid children, helping to reactivate maternal instincts of the Zetas. Later in the book we will discuss hybrid children and the women whose eggs have been used to create the hybrid.

As with previous star systems discussed, there are both positive and negative Zetas. The ones we have been discussing are the positive Zetas who had gone underground on the planet Apex. The negative Zetas, who helped create the chaos on Apex, eventually left the planet and settled in Sirius and Orion. The more negative Zetas are probably from the past. As we will see in another chapter, the negative Zetas do not want us or them to evolve out of the Third Dimension. On Earth, fear is a primary obstacle to growth, and Zetas are attracted to individuals who are fearful, thinking their DNA may help them recapture their emotions that had been bred out of them. However, at some level, the Zetas themselves are fearful, even if they are in denial, fearful of losing their dimensional physicality. They are hoping the new Zeta-Earth human hybrid will be unified and diverse and will be of unconditional love, hopefully leading them back to the Source.

CONCLUSION

After visiting our galactic ancestors, one can see that we have quite a dysfunctional galactic family. However, this is how we grow spiritually, to try to integrate the positive and negative. According to extraterrestrial sources, if we don't get it right in one life, we have the opportunity to get it right in following lives. The more knowledge we have to explain the bigger picture, the better understanding we have to cope with the microcosm. Humanity is evolving toward the Fourth and Fifth Dimensions, and most of our extraterrestrial ancestors are helping us get there. We also now understand the motives of those extraterrestrials who do not want us to evolve. Extraterrestrials have played a major role in humanity's development on Earth, and as we will see, they are responsible for the seeding of humanity. We are approaching the end of a major cycle, and hopefully humanity is in its last seeding as we evolve out of the Third Dimension into the Fourth and Fifth Dimensions.

Chapter Two

THE PLEIADIANS
Our Good Friends

As we examine the galactic civilizations in more depth, we find more information about the Pleiades than any other galactic civilization. The Pleiadians, in their desire to help Earth's evolvement, tried a daring experiment in Switzerland that would help put an end to all disbelief in UFOs and extraterrestrials. The experiment involved a man named Eduard Albert Meier, later to be known as Billy Meier. The purpose of the experiment was to see how humanity on Earth would react to the idea that we were not alone in the universe. They felt it was important that the people of Earth be awakened to the truth about the meaning of life. Earth had become spiritually stagnant and controlled by greedy men who caused wars to make money. Something needed to be done to wake up humanity.

The plan was to make physical contact with several humans and to teach them the truth about the physical and spiritual universes. The first contact was with a woman who was to carry the message. The Pleiadians gave her the information and asked her to go public, but she quickly gave up the idea after being ridiculed by family and friends. At that time, it was decided that humanity was too insecure in their spiritual development, and therefore, were not ready to take responsibility for themselves. The plan was postponed until humanity became ready.

An old soul from the Pleiades had agreed to incarnate on Earth to try to awaken the spirit of humankind. Eduard Meier was born in Bulack, Switzerland on February 3, 1937. Three others were born in Europe twenty minutes later that day in order to help with the mission. Two of them were later killed in a car accident, and a third remains anonymous.

When Eduard turned seven, the Pleiadians made the first contact with him in a Swiss meadow through a Pleiadian named Sfath. Eduard was taken aboard their craft and flown 40 miles up into orbit. Sfath was hoping to open up some old past life memories. Several years later, Billy (Eduard) was again taken aboard a space-craft and was told by Sfath to prepare for a very important mission. He was told to become aware of the language of spirit that is deeply written within all of us. It is a language of signs and symbols which allows one to interpret the meaning of telepathic messages that are sent to the spiritual side of oneself. Following several more years of telepathic education, Billy was told he would be left alone for eleven years to mature and contemplate what he had learned. During this period, Billy lost his left arm in a bus accident, an event that the Pleiadians had foretold.

For 300 years, the Pleiadians had occupied an underground base in Switzerland, and by 1965, 250 Pleiadians inhabited the base. They had been monitoring the thoughts of our planet's leaders and were studying the effect of "New Age" thought on mass consciousness. A Pleiadian woman named Semjase was chosen to be the physical contact with Billy. She had not been on Earth before and was not part of the underground complex. She lived on Erra, the home planet of the Pleiades. Semjase studied

Pleiadian Semjase

the German language spoken by Billy, and during the next ten years she studied the history of Earth and its people over the past several million years as she was preparing to educate Billy. During this time period, Billy had fallen in love with a Greek woman named Kaliope and they married. Billy had learned the art of meditation in the meantime and was able to connect with his spiritual self.

The first meeting date between Billy and Semjase was on January 28, 1975 at 1:00 p.m., when Billy received a telepathic message to get his camera and meet her at a remote location. From out of the clouds, Samjase quided her space-craft to the remote site. She had given Billy permission to photograph the craft to help awaken humanity. When Samjase exited the craft, Billy recognized her as a very beautiful woman with long red hair, quite normal looking. Speaking in German, Semjase grabbed Billy's hand and told him that the Pleiadians had previously tried physical contact with humans, but the people chosen were unwilling and lacked loyalty and sincerity. She said she would be meeting Billy again many times, and he could ask any questions he desired.

Billy was told that the Pleiadians had been coming to Earth for millennia and that they were not involved directly with any government or political leaders. Semjase emphasized that Pleiadians were not superhuman with supernatural powers. They were like the people of Earth, but had greater knowledge and wisdom, especially in technology and spiritual arenas. With this highly developed knowledge, they had developed telepathy and telekinesis and had access to higher realms of consciousness.

Semjase told Billy about the many life forms who travel the universe and sometimes visit Earth. Some of the extraterrestrials, if given the chance, would destroy whole planets and civilizations and force inhabitants into slavery. She warned Billy that some of these beings will be coming to Earth and there will be conflict. The Pleiadian mission was to warn humanity.

Semjase emphasized to Billy that it was time for humanity to learn about the true meaning of life and to live in peace and harmony with man and nature. She then made an interesting statement, saying that "God is simply a physical being, a governor, a human being who has evolved to a level of great understanding and knowledge through millions of lifetimes. Creation is the spiritual force that has the knowledge to create a universe. God is only a material being with knowledge of the human form and is subject to creation like all other forms. Never can a god take the role of Creation or control the destiny of a human." Pleiadians are no longer governed by gods or rulers, and now live completely free, enjoying their spiritual connection with Creation. The Creation itself never gives commands or demands worship in any fashion because it is an egoless, non-judgmental, spiritual force. She told Billy that Creation is eternal knowledge that quickens the growth of the human and that knowledge is never in need of commands and religion. The Pleiadians wanted to bring the truth of spirituality to the world. Billy then was told that there would be many more meetings, and that he was to pass on this important information as directed by the Pleiadian Council.

For notification of future contact, Semjase told Billy he would feel a cooling sensation on his forehead followed by a telepathic transmission of the meeting place. She wanted him to take many photographs and bring them to the public's attention. They met almost every week with Billy taking numerous photos. After the meetings, Billy would return home and type out their conversations. To aid with Billy's memory, Semjase arranged for a special computer-like device that would replay all that was said during the contact. The device was able to read the memory of the meeting that was stored in Semjase's mind and sent to the conscious mind of Billy, who typed the message. This usually took about two hours.

Billy would often take two or three rolls of pictures during the contact experience. Semjase wanted Billy to use the UFO pictures to give proof to humanity of extraterrestrial civilizations. Often she would situate the space-craft near various objects in the Swiss countryside to give perspective to the craft. The pictures began appearing in magazines and newspapers, resulting in a great disruption to Billy's life. The pressure began to build on Billy as he was trying to support a family when all this was going on. People were breaking into his home and going through his drawers. He was harassed, and in fact, there were several attempts on Billy's life. Needing to move away from his residence because of the disruptions, Billy found a nice farm in a remote area of Swizerland. By this time, he had accumulated over 1,000 pages of type-written notes which Semjase encouraged him to publish and speak about in public, which Billy found difficult to do. The contacts continued for three more years with more notes, astonishing pictures, metal samples, biological samples, crystals, and stones from other planets. Billy was also given permission to make a 30-minute film of the Beamship. On October 19, 1978, after 115 meetings, the Pleiadians ended contact.

One of America's top UFO investigators, retired Col. Wendell Stevens, wrote a book entitled *UFO - Contact from the Pleiades, an Investigative Report*. Lee and Britt Elders published a book of Billy's photographs, *UFO Contact from the Pleiades*. Billy published his notes in a set of books called *The Contact Notes*, but it was written in German. Randolph Winters, from California, made several trips to Switzerland over a three-year period to learn about the Billy Meier experiences. Billy had appointed him to be his spokesperson in America and Europe. Winters studied the notes Billy had taken and wrote a book titled *The Pleiadian Mission: A Time of Awareness*. Much of this chapter is based on Winters' writings.

LIFE IN THE PLEIADES

How do you tell friends that you spent your vacation in the Pleiades? Billy Meier had this dilemma after a three-day trip to the Pleiades with Quetzal, who had been the Pleiadian base commander on Earth for many years. Quetzal visited Billy a number of times after the normal contacts had ceased with Semjase. Billy was educated about the advanced civilization of the Pleiades with the help of Quetzal.

Location

The Pleiades contains a small cluster of six stars visible from Earth in the constellation of Taurus. The star system Taygeta is where the Pleiadians come from, located 500 light years from Earth. It takes our solar system 25,827.5 years to rotate around the Pleiades. Some of Earth's ancient architecture has been designed around the Pleiades. For example, the two base diagonals of the Great Pyramid are 12,913.75 pyramidal inches that add up to 25,827.5 pyramidal inches.

Erra is the home planet for the Pleiades in the star system Taygeta. Erra is ten percent smaller than Earth in a system that has eight other planets. A day on Erra is 23 hours 59.4 seconds, and a year is 364.35 days, with a 13- month calendar. Four of the eight planets are inhabited.

Physical Appearance

The Pleiadians are human and look just like us with several physical differences. Their skin is white, which is a natural function of evolution resulting in less pigment through multiple lifetimes. In stature, they are about the same size as Earth humans. The average life span is between 700 and 1,000 years. They are much older souls than Earthlings, in reference to the number of lifetimes and their advanced spiritual life. Because they live part of a second out of our time frame, when they come to Earth, they have to make a shift in their instrumentation in order to be seen.

Family and Marriage

Family life is different in the Pleiades when compared to Earth, with polygamy being the norm. Quetzal had a typical family, two wives and two sets of children. On Erra, one could love more than one person. Emotional problems such as jealousy had been eliminated on Erra, helping to explain how most Pleiadians could handle more than one marriage. The families live on the same land but in different houses, and the responsibility for raising children is shared by all. Divorce is not allowed because they feel it is an offense against the laws of Creation. Those who break the marriage laws are exiled from the planet, but this seldom happens.

Born in the Pleiades

Pleiadians fall in love, marry and have children. In this advanced civilization, reincarnation is an exact science, as it allows them to understand who is being born and their purpose in life. Spiritual leaders can read who they were in past lives and their purpose in the current life. Mothers try to find out what their new baby wants to be called, as a person's name is an expression of their spiritual development. After three weeks of pregnancy, the incoming soul will decide if it wants to inhabit the new body. Only during this time is an abortion allowed. Once the brain is developed, the spirit form transfers a copy of all its wisdom into the DNA of the brain where it will reside.

Early Life

During the first ten years of life, only positive influences on their environment are provided because this is a time for self-discovery that allows security and good behavior to help deal with negative forces. The next 70 years are spent in education, with up to 20 professions being learned. Following their education, Pleiadians will work four hours a day and for another four hours, they pursue creativity and spiritual growth, with an emphasis on the latter. Pleiadians take 100 percent responsibility for themselves and don't rely on god, myth, idol, or outside influences.

Communities

About 400 million Pleiadians live on Erra. They try to control their population in order to keep society spread out and to share the planet's resources equally. Quetzal told Billy that the Pleiadians had developed into a peaceful world of unity and harmony. Before the peaceful era, the Pleiadians had gone through thousands of years of wars and struggles.

Billy found the terrain of the Pleiades' planet Erra to be very similar to Earth's. Erra was mostly water with mountains, reminding him of Switzerland. He found no large cities with tall buildings. The small population of Erra was spread out into smaller communities. Most people lived in rural settings spread out from each other, living with nature. To preserve the environment of Erra, Quetzal said that most manufacturing and production was done on the other three inhabited planets in the Pleiades. Billy saw no smoke stacks and no pollution on Erra. The air on Erra was oxygen (32.4%) and nitrogen (67.3%) that Billy breathed. Housing on Erra was in the shape of domes, white in color. Organic intelligence, like that found on the space-craft, was built into the house to create a higher standard of living.

Diet and Health

The planet Erra has a green atmosphere, which contributes to good health and lack of stress. Pleaidians sleep only four hours a day, with sleep being deep and sound. Pleaidians do not eat as much as Earth humans and most are vegetarians because they know how food can affect thinking. On occasion they do eat meat, usually an animal that has been genetically engineered to look like a rabbit. Many Pleiadians have gardens, but do not actually work in them. Machines do the gardening, so Pleiadians are unaware of the joy of working in the soil and getting dirty. Food is grown in the ground, much like it is on Earth.

Thoughts

The Pleiadian society places a strong emphasis on proper thought. They understand the great power that a thought can play in one's life and, therefore, are protective of the consciousness of thought that is created by all who live there. Those who visit Erra with negative thoughts are asked to leave. Phones are non-existent on Erra because the ability to communicate by thought is practiced by most of the population. Pleiadians control their health by psychic balance, and they feel medical problems on Earth are caused by illogical thinking, in that thoughts create positive and negative changes in our cells.

Travel

Transportation on Erra is by tube, which is much like Earth's monorail system. People can travel around the planet in the tube, most of which is above ground, that is available to everyone. The tube can think and carry on conversations regarding one's destination, and it can even answer one's questions. Later we will describe space travel of remarkable speeds, with a journey to Earth taking only seven hours.

Government

After spending millennia in wars and conflicts, the Pleiadians realized that something needed to be done to save their civilization. They discovered a wise civilization in the Andromeda galaxy that was in the last stages of physical evolution and far more advanced than the Pleiadians. The Andromedans relied solely upon their spiritual consciousness for daily interaction. Following the Andromedan advice, the Pleiadians ended war and lived in peace. They formed a High Council comprised of the most spiritual beings who interpreted the wisdom from

Andromedan beings and then spread this information to all four of the inhabited planets. Today, before anything is passed, a high percentage of the voting public has to agree. The Pleiadians are involved in decisions and have a voting system much like our own.

Each of the four Pleiadian planets is a unitary-type system, subordinate to the High Council, whose central government is located on Erra. Comprising the Council are beings of human form who are half-spiritual and half-material with great knowledge and wisdom.

The Pleiadians are members of an Alliance of Civilizations that receive advice from Andromeda. There are thousands of these civilizations scattered around the Milky Way and Andromeda galaxies with a combined population of 127 billion people.

Economics

The Pleidians have a system of sharing their resources within their world and, as a result, have no economic system as we know it. There is no money or credit cards. Material possessions are all provided to them based on their contribution to the community. It is impossible to deceive or mislead someone because of their telepathic ability. In summary, everyone shares in the resources in accordance with their contributions without fear of greed and power.

Pleiades and Earth

Patrols from the Pleiades have been watching Earth develop for eons. They report on the many space travelers that visit Earth, about 3,000 ships a year. Only a small percentage are interested in making contact with humans. In 1976, the Pleiadians said there were seven other races that maintained bases on Earth. The Pleiadians abide by the policy of non-interference in the political and power structures on Earth. According to the Pleiadians, there are many extraterrestrial races who are not evolved and want to take us over and destroy us. The Pleiadians want to protect us and make their presence known to incoming ships through their network surrounding the planet. If we are invaded, it is not known if they will defend us.

Government officials on Earth are not trusted by the Pleiadians because those officials are mainly interested in power and control. Because of this attitude and threat of war against them, the Pleiadians feel it may be around 300 years before intragalactic contact is made.

THE SPACE-CRAFT

Mothership

To navigate the universe, the Pleiadians have two primary types of space-craft – the mothership and the Beamship. The mothership's dimensions are huge, measuring about ten miles in diameter in the shape of a large egg standing on end, with the bottom having three small round sections interconnected by huge braces. It is like a completely self-contained world that can support a population of 140,000 people. The inhabitants of the mothership enjoy it so much that they seldom return to the home planet. The ship is completely self-contained and even has gardens, lakes, and small mountains. To move about the ship, huge airways allow access to all levels. Essentially, they are floating pathways to move people up and down the craft. Hundreds of Pleiadians serve as crew members, assisting with control of the mothership. They are assisted by androids who were developed as helpers to handle most of the manual labor and technical work. They have been programmed with a highly-advanced intelligence, and their brains are organic matter grown by scientists. One advantage of androids is they stay alive for a long time. In addition, they are disease free and can be programmed for any kind of work, personality, and character. For spiritual reasons, the Pleiadians use androids rather than clones.

When visitors from other parts of the universe arrive on the mothership, they are given a special language converter to wear which handles communication. The device can read one's mind and pick up thoughts before they are spoken. From this information, it can create the words that need to be said. It is programmed with knowledge of thousands of different languages.

The mothership has the most advanced technology including the ability to travel in time, which some of the smaller Beamships do not have. Billy was told that a single second in timelessness can be equal to millions of years in normal space. Pleiadian scientists are in the early stages of time/space technology and do not understand it fully. The Pleiadians say the future is not a fixed thing but only a projection of events based on the present.

The Beamship

Beamships, which measure 21 feet in diameter and weigh 1.5 tons, require a crew of three. Most commonly, they are used for short trips with interplanetary capabilities. They come in many shapes and sizes. The name comes from one of the first drive systems they developed which relied on light-emitting devices to create power; hence the name. A Beamship has two drive systems, one for speeds up to the speed of light and one for speeds beyond the speed of light.

Most Beamships carry three people, with the three seats folding into sleeping couches. Some Beamships travel from one part of the galaxy to another and are a common way for Pleiadian citizens to travel. Billy was encouraged to take photos of the Beamship.

Instead of using explosion engines, the Pleiadians designed an implosion engine that causes matter to convert back into something useful. Energy goes out the bottom and is retained at the top of the craft. An energy field that surrounds the craft is blue in color and is used as a screen for protection from space debris. The screen also controls the line of vision, and pictures of the craft are possible only when the shield is dropped. The energy field that protects the ship allows the atmosphere to glide away instead of pushing against it, and at the same time it neutralizes the gravity force of the planet. The Beamship acts as a small planet itself with its own gravitational force. When flying, there is no movement inside the ship.

Some technology on board can be controlled by the mind. For example, the visual screen on the ship can be telepathically controlled. The screen can display information about any kind of life form such as birth time and the expected life duration of an individual. It can also interpret thoughts and feelings of people.

The technology aboard the Beamship can convert the ship and its passengers into fine matter particles that can travel faster than the speed of light in another dimension called hyperspace. This makes it possible to travel billions of miles in a second. The Pleiadians say that time is energy that cannot be seen. Time energy causes the rotation and movement of the Third Dimensional world. It controls the normal speed of all matter, and without time, space would stop moving.

In this other dimension called hyperspace, there are different kinds of energy particles that move at a much higher speed because time is different. Time does not exist in hyperspace the same as in the Third Dimension. The Pleiadian technology allows the Beamships to move into hyperspace and convert themselves into high speed particles, so once in hyperspace, the Beamship and its passengers are no longer in material form. After arriving at their destination, they reenter normal space and convert themselves back to original form. After a ship is 94 million miles away from the planetary orbit, it can enter hyperspace. This changeover into hyperspace creates a small tear in the time fabric so there has been a distortion in time. Sometimes dematerialization can pull objects other than the Beamship with it. A lot of energy is required to convert the solid matter of the ship and its passengers into fine matter to allow entrance into hyperspace. The 500 light years between the Pleiades and Earth can be covered in a millionth of a second. At 94 million miles from the Pleiades, the process is reversed and the passengers feel nothing. It takes another 3 1/2 hours to fly into the solar system of Taygeta. The technology to control the speed of particles while in a fine matter state is critical. If the speed of fine matter energy is too fast, one can be lost in the future; if too slow, one can be lost in the past. The Pleiadians have lost many of their forefathers in time.

The metal of the Beamship is a soft metal, a combination of material including lead that is found in the atmosphere of stars. It is congealed into a soft metal and tempered to make it hard. The final product is an alloy of copper, nickle, and silver, and some Beamships contain gold. Not all the elements of Beamships are found on Earth.

The Pleiadians have been able to put a special form of

intelligence into the cell structure of the metal. If the Beamship is separated from its owner, a special coding causes it to dissolve and break down. Billy was given a piece of metal, and within 24 hours it had deteriorated into dust. Computers on board the ship are organic and capable of highly intelligent thought. They perform most of the tasks on board the ship and protect the pilot when outside the ship.

Beamships have the technology to beam people on board using special matter converters, which break the body into fine matter and then reassemble it inside the ship. When Billy was beamed aboard the ship, it was located about 3,000 feet above him, and he would be aboard in an instant without discomfort. Eye witnesses have seen Billy disappear.

The Pleiadians also have small reconnaissance ships called telemeter ships that they use to gather information. Telemeter ships can be as large as nine feet in diameter or as small as a basketball. Telemeters can be directed by telepathy or by radio signals. Monitors in the ship can decipher all our major languages as well as our thought patterns that are reported by the telemeters.

PLEIADIAN SPIRITUALITY

The Pleiadians told Billy that we are moving into an era called the "Age of Aquarius," which is the era of spirituality. It will be an era of truth, enlightenment, freedom, wisdom, and harmony. The lessons will be learned through very difficult struggles. The Age of Aquarius will last 2,155 years and will be repeated every 25,860 Earth years. In this era, we will have a window of 800 years to achieve peace through spiritual growth and an understanding of Creation. If this opportunity is missed, we will fall into a deep, dark period that is controlled by negativity and illogical thinking. Civilizations historically last about 10,000 years, and ours is about 8,000 years old as we move into the most difficult phase. Spiritual people will see this as an opportunity for growth. However, most of humanity is caught up in the material struggles of life and will observe this period as a time of despair and chaos. The Pleiadians said the transition into Aquarius began on February 3, 1844 and by the year 2030, the transition will be complete. They believe the Age of Aquarius will be a time for accelerated spiritual awareness, not mass destruction. The era of ignorance is coming to an end, and we are entering a time of enlightenment. There will no longer be a need for misleading religious doctrine and religious leaders who claim to be mediators of God. Religion will no longer be used to control people and to prevent them from personal discovery. Most wars and loss of life have been based on religious differences. The Pleiadians say we are spiritual beings living in human bodies, here to gather wisdom from our material existence. First, we need the material senses to gather information and grow from the experience of a physical world.

The Pleiadians think of wisdom as knowledge that creates energy. The more one learns, the higher level of energy we provide to ourselves and Creation. Even if we are not aware of it, spiritual growth goes on in a natural process. Creation evolves by creating a material universe through which the soul learns. The spirit creates a material body, which goes through cycles of being awake and asleep, while it learns from the experiences. They believe the spiritual subconscious is an accumulation of all the wisdom learned from the many material lives we have lived so far.

Thoughts, according to the Pleiadians, are energy that have the ability to leave our minds and move out into space. They become part of a band of etheric energy circling the planet called the Akashic Record, which contains thoughts from all who have lived on Earth throughout history. These thoughts are stored in a logical way based on levels of evolution. Akashic Records use the knowledge of the past arranged in layers of memory, with each layer being a different level of evolution. Past knowledge is protected in the Akashic Record.

In order to help people on Earth, the Pleiadians use a form of telepathy to send information to thousands of people on Earth all the time. They send us ideas and visions that we can use. Usually this information is sent during our sleep. They do not want to hinder our free will or invade our thoughts, as they feel it is morally wrong. However, they use hundreds of small telemeters to monitor our thoughts and to broadcast thoughts for us to pick up. They seldom communicate by channeling, as they know it is too inaccurate.

On board a Pleiadian space-craft, they have an amazing instrument that can show a person's spiritual self on a screen. It contains knowledge about the time of birth, one's life purpose, karmic debt, past life trauma, and expected time of death. It can also read the feelings and thoughts of the conscious and subconscious mind.

Pleiadians view death as a transition in the endless path of spiritual growth on our evolutionary journey. Following death, our thoughts of fear, our emotions, hatred, anxiety, and other manmade feelings fade away.Our spirit is still conscious and is still aware of the material world. Spirit stays right here on Earth where it was originally born, because at our level of evolution, it is not possible to be born in another world or dimension.

Once we cross over to the other side, there is a time of rest and contemplation. There are no new experiences that we can gain from. The spirit is no more aware of the material world than we are of the spiritual world. After death, the average time on the other side is 152 years, but in this new age, some souls are returning in 15 years. There are three levels of consciousness one passes through on the other side before the soul is ready to return to the material world. If possible, souls have a tendency to stay within their own families. If spirit does move outside its family, it does cause a more difficult time during the next life. Souls do have the choice of gender upon their return.

Upon our return to the material world, our name reflects our level of evolution and path of learning. In spiritual language, our names have meaning. The spirit will try to communicate telepathically with the new parents about the name. Our soul knows how much accumulated knowledge it will bring into the new body. Wisdom from past lives will be carried forward into our new life. This is why we all have different IQs and aptitudes. After the formation of the brain in the fetus, the spirit is able to make a copy of its wisdom and implant it into the DNA of the brain for the material body to use. The new physical body of a soul is the product of the DNA of the mother and father or family genetics. However, the spirit is reflected in one's face. If one could look at pictures from past lives, the face would look somewhat similar, especially around the eyes.

Upon birth, we will have a new consciousness to experience life. Our subconscious is empty except for the memory inside the womb. We enter this world with a clear psychology that has no emotions of hate, anger, or prejudice, as we are ready to start another life journey with a clean slate.

THE SECRETS OF EARTH

During Billy's many contacts, he was given permission to ask any questions he wanted, and he did ask many, including ones about the mysteries of Earth and life. In this section we will highlight a few of Earth's mysteries.

The Bermuda Triangle is part of the old continent of Atlantis, where descendants of Atlantis still live peacefully far beneath the ocean. A dimensional door to a parallel universe exists in the Bermuda Triangle that explains many of its mysteries. Billy was told that the dimensional door is caused by two giant suns located 720 light years from Earth. High energy radiation from the two bodies come together at certain points on Earth, causing a rip in the time fabric. Two other locations exist where this energy hits Earth – off the coast of Japan and near Madagascar. The crossing energy opens a dimensional door to the parallel universe. There are three parallel Earths – a prehistoric Earth, a newly-formed Earth, and an Earth 500 years in the future.

Billy was told about a race of extraterrestrials called the Hyperboreans, who live under Mount Shasta. The main entrance to their civilization is on the east side of Mount Shasta. The Hyperboreans' forefathers came to Earth 30,000 years ago.

The statues on Easter Island are connected to the city Tiahuanaco, just south of Lake Titicaca in Bolivia that was inhabited by a pre-Incan culture that existed 300 B.C. to 900 A.D. It represented the last colonization by extraterrestrials around 13,000 years ago. Out of this group came Viracocha, the most famous South American god who conquered Tiahuanaco and settled on a small island near Easter Island. He and his people were Lyrans with giant bodies, who eventually developed an illness and had to leave Earth by space-craft. They kept in contact with a race of humans in Andromeda who came to Earth 2,568 years ago, stayed for 20-plus years and constructed the Nazca lines. The Andromedans were unable to acclimate to Earth and had to leave as well.

The Greek gods, such as the Titans, were also extraterrestrials who colonized the Earth. The Titans were descendants of the ancient Lyrans, who were giants. Hercules was a Greek god who stood over ten feet tall and was a descendant of the Hyperboreans. Billy was told that some of the great Sumerian gods stood over 24 feet high. Both humans and the Pleiadians are descendants of Lyrans, confirmed by other sources.

The Pleiadians said that Jesus survived the crucifixion and lived in Kashmir where he was married and had children. On the way to Kashmir, his mother Mary died. Jesus' son had returned to Israel and deposited the corrected history about Jesus in a cave. It was discovered last century by a Lebanese Catholic priest who was able to translate the Aramic writing. He had been directed to the cave by the Pleiadians. Much of the text was translated before the priest was killed. Billy Meier was in possession of the manuscript and did have it published. (This story is described in detail in the author's book entitled *The Unknown Life of Jesus: Correcting the Church Myth*).

The Pleiadians also confirm that the Great Pyramids were built by extraterrestrials. The first great Pyramid was built by Lyran ancestors about 73,475 years ago. (Other sources say the Sirians ,who were ancestors of the Lyrans, built the pyramids.) They also confirm that it housed an entrance to Inner Earth. They talked about a group of extraterrestrials called Bafath, who secretly came to the Giza Plateau. For 2,000 years they have tried to rule Earth and actually have controlled many religious leaders. The Pleiadians intervened and relocated them to another planet, but they have returned. Through the Thule family in Germany, the Bafath seized control of Adolph Hitler, who was convinced he was doing the right thing for Earth.

In 1975, the Pleiadians claim there were over 17,000 people on Earth who were in contact with the Pleiades, but only through thought. At that time, no government was in regular contact with extraterrestrials. The Pleiadians emphasized they have a policy of non-interference with political and power structures on Earth. Our Pleiadian friends are very concerned about our welfare and view us as somewhat of an insane society. They had gone through an era of wars much like what Earth is going through now, but found a plan for peace that worked for them. Understanding that Earth is on a path of evolution, they want us to avoid the mistakes of past civilizations and are here to guide us through these difficult times.

Chapter Three

THE SIRIANS
A Serious Fish Tale

The Sirians left many clues regarding their influence on the planet Earth. One has only to examine the legends of a primitive African tribe called the Dogons, who have preserved this knowledge through the millennia about our Sirian visitors. The Sirians also had a great effect on the Egyptian civilization, as witnessed through legend, architecture, and calendar. Their influence was also seen in the ancient civilizations of Lemuria and Atlantis, and continued with our civilization when ancient Sumer arrived on the scene in Mesopotamia. The Sirians have been our guardians, guides, and overseers, and science has shed some light on this intriguing galactic civilization.

The Dogons

Robert Temple is a well respected British scientist who is a member of the Royal Astronomical Society, the Royal Historical Society, and the Egypt Exploration Society. This respected researcher stunned the scientific community when he published a book entitled *The Sirius Mystery: New Scientific Evidence of Alien Contact 5,000 Years Ago*. His first book, published in 1976, claimed the planet Earth was visited by intelligent beings from the star system Sirius. The book became a best seller and was reviewed by the prestigious scientific journal *Nature* and by *Time* magazine. Not everybody wanted this book published. Temple received extreme hostility from certain intelligence agencies in the United States. They planted false information into his security file and Temple was blackballed by some organizations. In the 1970s, the CIA actually stole a manuscript from him. For three weeks, the CIA tried to persuade, and ultimately convinced, a producer not to produce a video of the book. For 15 years,

Temple was persecuted by the intelligence agencies, including the Soviets, who wanted to suppress any study of extraterrestrial intelligence. Even NASA got into the act by criticizing the book. Later, we will discuss in more detail how the Secret Government wants to keep information suppressed about extraterrestrials.

Temple's research involved the Dogon tribe located in Mali, Africa, a sub-Saharan state whose nearest city is Timbuctoo. The tribe had originally been located on the southern bank of the Upper Niger River. Two anthropologists, Marcel Griaule and Germaine Dieterlan, made the first important discovery connecting Sirius with the Dogons. The anthropologists lived with the Dogon tribe for a period of time doing research and managed to gain the tribe's trust, resulting in shared secrets by the head priests. Prior to the sharing of secrets, the priests held a tribal meeting and a policy decision was made to share the secrets. One secret was that all forms of matter are emitted from Sirius. They also told them, "The starting point of Creation is the star which revolves around Sirius and is actually named the 'Digitaria Star.' It is regarded by the Dogons as the smallest and heaviest of all the stars. It contains the germs of all things. Its movement on its own axis and around Sirius upholds all Creation in space. Its orbit determines the calendar." This esoteric knowledge had been handed down orally for centuries in unbroken chains and carefully guarded to prevent censorship. An example of the deep trust they showed for the anthropologists happened in 1956 when Marcel Griaule died and a quarter-million tribal-people gathered for his funeral in Mali. They had come to revere him as a sage, equivalent to their high priests.

The Dogon tradition had been traced back to ancient Egypt, which revealed a contact in the distant past between Earth and an advanced race of intelligent beings that had come from another planetary system several light years away. Between 7,000 and 10,000 years ago, our planet had been contacted by the Sirians.

The Dogon Knowledge

The brightest star in the heavens is Sirius, not Venus or Jupiter because they are planets. Sirius is bright because it is large and close, bigger than the sun and other nearby stars. It was discovered that every 50 years, a little star called Sirius B revolved around Sirius A, the larger star. This gave the appearance that Sirius was wobbling. Sirius B is a dwarf star, a star that is feeble, yet strong, not emitting much light but possessing a great gravitational force. An Earth human on Sirius B would be only a fraction of an inch tall and weigh thousands of pounds because of the gravity. The first photograph of Sirius B was taken in the 1970s. Some astronomers had been predicting a third star, which would be called Sirius C, that the Dogons had known about for millennia. In 1995, Sirius C, predicted in Temple's 1976 book, was confirmed by astronomers.

The two German anthropologists concluded that the Dogons had such advanced knowledge about the Sirians, possible only through direct contact. The Dogons called Sirius B "potolo," with Tolo meaning star and Po a cereal grain (Digitare Exglis) in West Africa. The Dogons referred to Sirius B as Digitaria and claimed that it had an elliptical orbit around Sirius A. This obeys Kepler's law of planetary motion. How did the Dogons know this?

The Dogons also knew that the orbital period of Sirius B was 50 years around Sirius A, later confirmed by science. Legends of the Dogons also claim that Sirius B rotated on its axis, another fact confirmed by science. The legend states, "As well as its movement in space, Digitaria (Sirius B) also revolves upon itself over the period of one year and this revolution is honored during the celebration of "Badorite." The day of Bado, according to the Dogons, is when a beam of light carrying important signals strikes the Earth from Sirius B. Dogons also described Sirius B as being tiny but very heavy, again confirmed by scientists who say that Sirius B is the tiniest form of a visible star known in the universe and also the heaviest.

Another claim this primitive tribe made was the existence of a third star in the Sirius star system they call "emma ya," claiming it traveled along a greater trajectory in the same direction and at the same time (50 years) – but it is four times as light, when comparing it to Sirius B. As mentioned, astronomers discovered Sirius C in 1995.

Dogons had other knowledge about the universe, revealed by their drawings of Saturn's rings. They knew planets revolved around the sun, and they knew Jupiter followed Venus by turning slowly around the sun. They were aware of the four major "Galilean" moons of Jupiter, the moons Galileo had discovered in 1609 with a telescope. They described the Milky Way of our galaxy as spinning in a spiral, and they described Earth as spinning on its axis. Where did this information come from? The Dogons say from the Nommo.

Nommo is the collective name for the great cultural hero and founder of civilization who came from the Sirius star system to establish civilization on Earth. Nommos were amphibious creatures that were represented in tribal drawings. Dogons referred to the Sirius star system as the land of fish. Landing on Earth by the Nommo is called "the day of the fish," and the planet where the Nommo came from is in the Sirian star system.

The Dogon often say that in order to atone for our impurity, the Nommo dies and is resurrected, which acts as a sacrifice for us in order to purify and cleanse the Earth. Their legends say the Nommo will come again, and a certain star in the sky will appear once more and will be the testament to Nommo's resurrection.

The Dogons center their life and religion around the Sirius star system, not our solar system. They honor the visitation to Earth by the amphibians (Sirians) who brought us civilization. One of their four calendars includes a Sirian calendar.

From the legends of the Dogon, Robert Temple concluded that the Sirians had visited Earth, giving knowledge to Earth humans and starting civilization.

The Nommos

As mentioned, Nommo is the name given to the extraterrestrials (Sirians) who visited the ancient Dogon tribe. According to the their myths, the Nommos gave all their life principles to humans. The Dogons also have drawings of Nommos to give them an idea what they looked like. The name "Nommo" originated from the root word "nomo," which means "to make one drink." Legends say they descended in a spinning ark from the sky, and that Amma, the god of the universe, had sent the amphibian Nommos to Earth. Collectively, the Nommos were known as "masters of the water" whose purpose was to monitor and teach. Dogon legends say the ark landed on Earth, northeast of today's Mali Africa, where the Dogon tribe originated. The tribe heard a vibrating sound of thunder as the ark landed, while another ark hovered in the sky at a great distance. Their legends say the Nommos will come again and there will be a resurrection of the Nommo, who are considered the father of mankind and guardian of spiritual principles.

After the ark landed, a series of interesting events took place. Amphibious creatures were inside the craft. Something described as a quadruped appeared from the craft and pulled the ark with ropes to a hollow of water, and the ark floated like a huge dugout canoe. In the center of the ark, the Nommo stood, and he was "O Nommo" or Nommo of the pond.

Four types of Nommos are discussed in Dogon myths. The Nommo Titiyayne is the messenger of the Nommo Die who executes the message. The Nommos who come to Earth in their spaceships are thought to be in this class. The ONommo class are the "Nommo of the pond," and "will be sacrificed for the purification and reorganization of the universe . . . then he will rise in human form and descend on Earth in an ark, with the ancestors of men . . . then he will take on the original form, will rise from the water and will give birth to many descendants," according to Robert Temple. He says this suggests that a group of Nommos from the Nommo Titygayne class volunteered to be reincarnated as humans during a period of official absence from Earth. The naughty disrupter Nommo is called the Ogo, or Nommo Anagonno. Legend says Ogo was about to be finished during Creation, but rebelled against the Creator and introduced disorder in the universe. Ogo will become the Pale Fox, according to the Dogons, and be the image of the fall, similar to Set in Egyptian mythology.

The Nommos needed to breathe our atmosphere, which the Dogons said they did through their clavicle. Some researchers think Nommos were much like dolphins, except with arms and legs. The blow hole may have consisted of two slits. The tradition of the mermaids may describe the Nommo, with the upper torso having arms and hands. There is a sea creature called the "dugona," which looks like a small manatee but with a tail exactly like a Dogon Nommo. They were common in the Persain Gulf just off the coast of Sumer in present day Iraq, but are now found near the Great Barrier Reef.

Similarities have been found between western spirituality, including Christianity, and the Dogon religion. The tribe members worship Amma, a supreme god who created the universe. They also worship a crucified and resurrected savior from Sirius. Theosophist writer Alice Baily maintains that the "Great White Lodge" of Freemasonry was based in the Sirius star system. She says that Sirius is always beaming helpful rays to the people of Earth, who live in ignorance, violence, and oppression. Because of Earth's violent nature, the Sirians have tried to civilize Earth without much success. Freemasonry was one method that has been tried to civilize us. Alice Baily wrote, "One great fact to be borne in mind is that the initiates of the planet or of the solar system are but the preparatory initiation of admission into the great lodge on Sirius." She also wrote, "in the secrets of the sun Sirius are hidden the facts of our cosmic evolution and incidentally of our solar system."

The Egyptian Connection

The ancient Egyptian name for Sirius was Sothis, and many parts of Egyptian life were centered around Sothis. Even the ancient Egyptian calendar was based on the movement of Sirius. Following 70 days in the Duat (underworld) was the helical rising of Sirius (in conjunction with the rising sun), which was the first appearance of Sirius on the western horizon just before the rise of the sun. This event occurred once a year giving rise to the Sothic calendar and Egyptian New Year, announcing the inundation of the Nile, occurring during the prime of summer. Helical Sirius was important to both ancient Egypt and Dogons. Gigantic temples were constructed with their main aisles orientated precisely toward the location on the horizon where Sirius would appear on its helical rising. The light from Sirius was channeled to focus on the altar in the inner sanctum. It shone into the Temple on New Year's Day and intermingled with the light of father RA (ancient name for the sun) on the horizon.

Sirius is located in the constellation of Canis, or dog, and is often referred to as the "dog star." Legends often called Sirius the red star because it was red during its helical rising. Many references in Latin literature talk about the "dog days of summer," which follow the helical rising of Sirius in the summer. Sothis (Sirius) was identified with the goddess ruler Isis who was head of the Egyptian pantheon, and Osiris, the husband of Isis, who was identified with the constellation Orion. Isis was the greatest goddess of all Egypt during the era of 4,000 B.C. to 3,000 B.C. Her duties were often involved with the dead, and Osiris, companion to Isis whose soul dwelt in Orion, was also god of the dead. In early Egypt, heaven was usually associated with Sirius, described as being prolific with vegetation and water.

Astronomically, Sirius was the foundation of the entire Egyptian religious system. Isis and Osiris had been sent to Earth to help give primitive humankind the art of civilization. They also taught men how to care for the dead. The god Anubis was involved with embalming Sothis (Sirius) for the 70 days in the Duat, when Sirius was not visible. Anubis is symbolized by the dog or jackel head, whose father was Osiris and mother was Nephthys, sister of Isis and goddess of the invisible. An embalmed mummy is suppose to come alive again, as symbolized by the mummy of Sothis coming alive during its helical rising.

Amphibian Legends

Other countries have legends about the importance of Sirius, including Babylonia, which teaches about the god Oannes, the fish-tailed amphibian who came from the heavens and founded civilization on Earth. He was the teacher of early man in all knowledge. The amphibian Oannes who brought civilization to Earth is often equated with the Sumerian god Enki (EA). According to Sumerian writings, Enki is the god responsible for early civilization and for the genetic creation of man as we know it today. Enki is a god who slept at the bottom of a watery abyss, similar to Oannes who retired to the sea at night.

In 290 B.C., Berossus, a priest of the god Bel, wrote a book entitled *Babylonian History*, claiming that a group of amphibious beings founded their civilization. This group of amphibians is called Oannes, becoming later the fish-god of the

Philistines known as Dagon. Eventually Oannes/Dagon became an agricultural deity. This coincides with the Dogon legends about civilization being founded by amphibians with fishtails from Sirius. From their description, it seems as if the amphibians were a cross between a man and a dolphin. Berossus described the amphibians as such: "The whole of the animal was like that of a fish; and had under a fish's head another head, and also feet below, similar to those of a man, subjoined to a fish's tail. His voice too, and language, was articulate and human, and a representative of him is preserved today. When the sun set, it was the custom of this being to plunge again into the sea, and abide all night in the deep, for he was an amphibian."

Giving credence to the amphibious legend is Plutarch, the early Greek historian, who wrote, "Moreover, Eudoxus says that the Egyptians have a mystical tradition in regard to Zeus, that because his legs were grown together, he was not able to walk." Again, this is very similar to the amphibious Oannes of the Babylonians, who had a tail for swimming instead of legs for walking.

The Greeks also had legends regarding amphibious fishtailed beings with human-like bodies. Cecrops was an ancient amphibian who founded Athens and became the first king prior to the time goddess Athena appeared on the scene. As time went on, the fishtail became more serpent-like. Historian Diodorus Siculus, in the first century B.C., wrote that the Cecrops originated from Egypt.

Triton was another Greek amphibian with a fishtail and man's body. Triton was the name of the Oracle Octave Center at Lake Triton in today's Libya, the birthplace of Athena. When the Aryans arrived in India, they took with them the legend of a water god named Trita who can be found in the earliest Sanskrit texts, the Vedas. Triton's name meant "one third," and the Sumerian god Enki's name Shanabi literally means "two thirds." In the famous Sumerian legend regarding the Sumerian king Gilgamesh, he was described as being two thirds god and one third man. The legends of Triton/Trita seem to preserve the Summerian/Babylonia mythology.

Typhon, another legendary figure of great importance, had

a human top-half body and a dragon or serpent lower half. His temper was terrible. According to Egyptian mythology, Set, also called Typhoon whose brother was Osiris, was god of the storm and of gloom.

Some ancient art shows the goddess Isis as half-human and half-fish. Her husband Serapic (another name for Orion) is also portrayed as fishtailed.

Even the Chinese have legends about their civilization being founded by an amphibious being named Fuxi, who had a man's head and a fishtail. The date 3322 B.C. was ascribed to Fuxi, who invented the system of the trigrams and hexagrams in the *Book of Change.* The information was revealed to him by another amphibian who rose up out of the Yellow River and had the symbols displayed upon his back. Fuxi and his wife Nu Gua repaired the broken heavens and were founders of civilization after the Great Flood.

Much evidence has been presented to show that the Sirians probably visited Earth during ancient times according to legends and myths, along with circumstantial evidence found in Egyptian architecture. Several individuals have been contacted directly by the Sirians, both through physical and telepathic means. We will now look at contemporary evidence to see if there is any confirmation of these ancient legends.

ACCORDING TO THE SIRIANS

Because of the law of non-interference, enlightened extraterrestrials are not allowed to overtly interfere with a planetary system until the planet and its people have evolved to a certain level. However, they are allowed to guide us in subtle ways, and this section of the chapter will discuss the information Sheldon Nidle and Virginia Essene have received from the Sirians in their book *You are Becoming a Galactic Human.* The Galactic Federation, according to Sheldon, had appointed him as a representative between Earth and Sirius. Similar to Billy Meier, Sheldon at a young age was instructed by an extraterrestrial, this time from a being from Sirius named Washta. His experience began one night as a child when he was visited by Washta, who told him that he was from a star system called Sirius B and that he was to be Sheldon's instructor. Confirming his extraterrestrial roots, he had Sheldon look out the bedroom window to see seven UFOs aligned in a V formation that he then instructed to change formation and they did. This was to give proof that the messages Sheldon was going to be receiving were real. Sheldon's younger sister witnessed the demonstration as well, and later both he and his sister visited the mothership in the form of an out-of-body experience. Washta reappeared again at a later time and told Sheldon about wars in space, about major catastrophes predicted for Earth that the Sirians would not allow to happen. Following these early childhood experiences, four of the six members of the Sirian Council began imparting knowledge to Sheldon though mental telepathy.

Virginia Essene also had the gift to communicate mentally with these Sirians through telepathy. Sheldon and Virginia were receiving a lot of the same information, so they decided to collaborate and write the book together. Their information confirms other information received from extraterrestrials plus new information, whose purpose according to the Sirian Council was to introduce humans to the Galactic Federation, so the supreme creative force could be brought into our planet. The second reason they are here is to help Earth humans to acquire full consciousness once again. They are here to oversee and assist us in our evolution. They said the Sirians and ascended masters have come to Earth to create a new love and wisdom across the entire galaxy, and that it is time for Earth to belong to the galactic family.

Before we get into the information shared by the Sirians, it might be wise to tell the reader about Sirius. Sirius is located about 8.7 light years from Earth and is comprised of nine stars with three major stars, Sirius A, B, and C, with Sirius A being the larger of the three. Sirius A is two and one half times the mass of our sun and 10,000 times brighter than Sirus B. Interestingly, if Sirius B were on its own, it would be the brightest star in the sky. Sirius B is a blue-white star that has a radius of 0.0078 times our sun and takes 50 years to rotate around Sirius A. Sirius C is a small red dwarf star whose planets are covered with a vaporous layer of clouds. Researcher Robert Temple believes Earth and Sirius have a shared movement in relation to the galaxy, believing they are in continuous resonance with each other.

The planet of Sirius B is one and one fourth times larger than Earth, which includes one large ocean about the size of the Pacific Ocean. Huge mountains rise near the seacoast ranging in heights from 3,000 to 11,000 feet, with the interior of the continent being broad hilly plains. A solar year is 440 days, and the planet has a calendar of 10 months of 44 days each. The climate on Sirius B planet is considered semitropical with an average temperature in the mid- to upper- 70 degrees Fahrenheit range. There is a blue hue to the sky. Large rivers meander through the plains, connected with a large system of lakes. The soil has an orange and purple tinge, mixed with brown, with many trees manifesting blue bark. The planet has a lot of natural beauty with abundant life energy. The Sirians agree they have a beautiful planet and take much pride in it.

Sirius A has the original nonhuman inhabitants while Sirius B is populated by galactic humans. The *Voyager Series* of Anna Hayes* claims that the Anunnaki extraterrestrials that are written about in Sumerian cuneiform tablets are from Sirius A. Sirius C and D are used mainly for administration tasks and storage. Three planets are inhabited by nonhuman creatures. These beings are fully conscious and are similar to a seven- to eight-foot lion. They are humanoid in appearance and have cat-like fur and a lion-like face.

There are several types of humanoid races living in the star system, with one race having a pale skin color and the other a light blue color. The height of males is around seven feet and females about five and one-half feet in both races. Their eye and hair color vary in color, and they are slender in appearance.

The Cetaceans

The Sirians and other extraterrestrials have told various sources that most of the existing life and vegetation on Earth originated from other star systems, especially Vega in the beginning. The cuneiform writings of ancient Sumer also confirm this. The Sirians told Sheldon and Virginia there have

* Anna Hayes is now writing under the name of Ashayana Deane

been wars on Earth, including nuclear, that have destroyed life forms and the extraterrestrials have had to reseed the earth. The Sirians were responsible for bringing the whales and dolphins to our planet, collectively called the cetaceans. It took about two million years for them to develop into what they are today. They were able to change from being on land to being in the sea, and in the beginning they were almost etheric. Pre-cetaceans were about seven feet tall and bipeds with fur and a small stump for a tail. The Sirians were responsible for the development of cetaceans who now serve as the guardians of our planet.

Whales, according to the Sirians, have the task of preserving life in the ocean and on land through their higher consciousness. Once humans become fully conscious beings, they will take over that task. Dolphins and porpoises help species in the oceans by maintaining their biofield energies. They can also assist specific individuals and species on land. Whales are involved with the greater picture, while dolphins are more involved with the specific.

The Photon Belt

Photon energy is a force found in the Third and Fourth Dimensions and comprises the photon belt which gives off an enormous amount of gamma rays and other radiation particles. The Sirians had to create a hologram in 1972 that encompasses Earth in response to the actions of negative extraterrestrials, who had altered the sun's polarity in a negative way. The hologram was able to reverse the sun's polarity that occurred in 1987 and 1989 in preparation for the photon belt. The actual entrance into the photon belt by Earth will not be known until two to six months before it actually occurs. Those who enter the photon belt will ascend into new realities and dimensions.

Once Earth enters the photon belt, there is a chance of major natural catastrophes occurring. The Sirian Regional Council wants to protect the Earth from these catastrophes as does the spiritual hierarchy on Earth because of humanity's raised consciousness. The cetaceans have also petitioned the Galactic Federation not to allow these catastrophes. To avoid a catastrophe, the Sirians can activate the sun's energy portals.

By analyzing the degrees of imbalance, they can induce a high energy frequency pulse to help rectify the imbalance and prevent the disasters.

On the other hand, the dark forces do not want humanity to go through this transformation and are interfering with the sun's energy. The purpose of the planetary hologram is to allow us to see reality as unchanged, and to allow us to gradually move through the change and shift in reality. The Sirians are concerned about the negative Zetas who are being used by the dark force, the Dracos Reptilians. The Zetas are attempting to create a new type of solar system that would allow their race and reality to become permanent, thus ending any intervention by galactic extraterrestrials in our solar system.

Presently, every person on Earth is being altered genetically in various degrees within the morphogenetic fields to help bring this new civilization into reality. The Sirians say that Earth is a planet of grace, and each human must realize he or she came to Earth to put love, light, and forgiveness around the entire planet. The amount that our consciousness is raised will determine the timing of a Sirian landing and first contact, which will be determined by the Galactic Federation.

The photon belt will completely alter the physical body. Humans will be changed from a dense physical body to a semi-etheric body. After the transformation, our bodies will be able to rejuvenate themselves and become virtually ageless. The semi-etheric body will respond in many ways like a thought form, and our new bodies will change as our thoughts change. How can this happen?

We come back to our DNA. All humans have two strands of DNA, but before our "fall" into the third-dimensional material world, we had full consciousness and twelve strands of DNA. The Sirians are helping to increase our DNA from two to twelve strands. Our cells will then be able to easily interact with the interdimensional spirit that each entity possesses. Presently, every human, whether conscious or not, is undergoing a great deal of genetic alteration that is preparing humans for transformation and ascension..

This new DNA configuration will follow the shape of the Star of David. Each of the six portal points will interconnect with

Third Dimensional reality through Eighth Dimensional reality of a multidimensional universe intersect. The cells will possess multidimensional scalar wave antennae that can easily pick up and process any important messages that the cell and its DNA package are given by the soul. Scalar waves are non-hertzian wave forms with the ability to propagate themselves in any form of multidimensional reality and carry information as they travel. Scalar waves form the basis of all the various energies of creation. It is where light comes from, and scalar waves make all things in creation. These waves are timeless energy and based on pulse, the pulse of time.

A new chakra system will develop that will evolve from the present seven chakra system to thirteen. The 12th and 13th chakras will be the final chakras and will be called the galactic male and galactic female energy centers, because they contain the ideal female and male prototype. As a result of the new chakra system, many segments of the brain that have atrophied because of the Atlantean genetic experiments will be returned to their former size and shape. This will allow humans to regain their psychic ability, telepathy, telekinesis, clairvoyance, and clairaudience. Our Sirian friends will help teach us how to communicate with our personal consciousness. We will have to practice self-control and neutral imaging so we will not harm ourselves or others. We will be moving from a Three Dimensional reality to a Fifth Dimensional semi-theric reality. There will be a new world where all minds share a common consciousness, and Earth humans will no longer have a separate world. Almost all Sirians have developed the 13-chakra system and have a fully conscious concept of reality.

The Galactic Federation wants to educate us about the new reality before the photon belt arrives. We need to learn to regulate our new conscious process and to understand how to control and utilize our thought form. We will also have the ability to communicate properly with those who have died. Our bodies will be lighter, and we can use our thought forms to rejuvenate and overcome aging. We will evolve into both a being of spirit and a physical being. A new age will have arrived, made possible by Earth's spiritual hierarchy and the Galactic Federation.

The star tetrahedron plays a key role with interdimensions and is related to the interdimensional human. It represents the great light body that surrounds the cellular structure, and is the intermediary between interdimensional portals and interdimensional realities. The star tetrahedron is the overlap around which all DNA is actually processed. Energy is passed up through the light body star tetrahedron and is transformed into various interdimensional portals. The information is then transferred back from the higher dimension through the light body star tetrahedron to the physical Star of David and the six-pair configuration of the DNA. The star tetrahedron represents a Fourth Dimensional and Fifth Dimensional vibration which has the capability of being utilized in the Third Dimension.

Sirian Spirituality

Sirians are Third Dimensional beings with Fourth and Fifth Dimensional reality. They have complete control of physicality and can transform their bodies to an interdimensional light body. By increasing their frequency, Sirians can disappear in front of a human. Approximately one-half of our galaxy has moved into the light.

Like most enlightened extraterrestrials, the Sirians believe in a spiritual hierarchy composed of angels, archangels, and ascended masters, who serve as spiritual mediators through the Eight-Dimensional evolutionary energies. The hierarchy transfers these energies through the appropriate interdimensional portals for transport into our Three-Dimensional energy patterns that create the physical universe. Each physical universe has its cycles and its own pattern and was created to show how light can transmute darkness to produce an even higher light, composed of the highest love. The Sirians believe this love will transform our galaxy toward the Supreme Creative Force, whose chief emissaries are the Time Lords.

According to the Sirians, the Time Lords are the divine shepherds of physical creation, and time is the unit that controls all of physical creation. The Time Lord's task is to regulate and oversee the physical creation according to divine right action. Their purpose is twofold: 1. They had to create the eight dimensions of physical creation. 2. They aid the spiritual hierarchy in regulating the spiritual energy exchange between each dimension. The spiritual hierarchy controls interdimensional energy transfers by use of interdimensional gates called star gates, which serve as flux barriers between dimensions and regulate energy flow, allowing sufficient energy exchange to successfully maintain the viability of physical creation. It was the Time Lords who constructed the Third Dimension galaxy according to the divine plan. Both Time Lords and the spiritual hierarchy transmute Creation into a new reality.

All the spiritual systems are organized from one to eight dimensions. After Dimension Eight, there are a number of higher spiritual dimension that deal with the nature of the Godhead (Time Lords and Archangels). Dimensions One through Seven were created to experience various aspects of the physical creation. Dimension Eight is where all these lower level qualities come from. Above Dimension Eight sits the infinite Godhead and the realms of the Time Lords and Archangels.

The lower dimensions are defined as follows:

Third Dimension - *Physical is reality*
Fourth Dimension - *Physical is not quite reality*
Fifth Dimension - *Form's purpose is defined*
Sixth Dimension - *Forms develop a purpose*
Seventh Dimension - *Spirit has form*
Eighth Dimension - *Light becomes spirit*

The purpose of the angels and the Archangels is to receive and distribute the love and light of Creation throughout the physical universe.

The Sirians say there is a Time Lord for every part of a soul and for every aspect of a soul. A soul force may be divided into many individual beings. The Time Lord is infinite because time is non sequential.

The spiritual hierarchy needs physical guardians in the physical realm who can perform the task of energy exchange on a smaller and localized scale, such as a star system or planet. The guardians of any planet must be both physical and spiritual, and these beings must possess full consciousness. A guardian brings in the creative energy and regulates this energy for the planetary life sphere. There are two major species designed as a guardian group for Earth: the cetaceans (whales and dolphins), and Earth humans, who were originally brought from other star systems to Earth. The guardians help Earth and all its life forms to flourish, and guardianship means raising consciousness regarding the environment and relations with others.

Being of full consciousness, the Sirians are aware of the spiritual laws and concept of karma. They know that karma is the great collective need that each soul has to become in order to be a fully ascended being. Each soul provides a piece of the puzzle that society requires for its growth and security. By selecting a path that allows both growth and service, it will be possible for anyone to advance spiritually without a need to incarnate again to enter galactic presence. When an individual has transcended the physical, and when the angelic judges decide to promote him or her to a full spiritual being or spiritual counselor, the individual can join the full spiritual hierarchy or the ascended masters.

The Sirians say that the energy pattern of Earth human souls is basically the same as Sirians, but the main difference is that Sirians are in full consciousness with their light bodies. Earth humans have been genetically manipulated to be disconnected from the love and light power. In other words, they are separated consciously from their Higher Self.

The Laws of Society

The Sirians try to live by the four laws of society that act as a guide for their development. These laws are based on two main principles. The first principle involves the importance of personal growth in an individual. Growth involves raising one's consciousness, a condition that is accomplished only by fully exploring one's Higher Self and by being of service to others. The second principle is that each person's soul light shines on everyone in a unique way, and each soul light contains a piece of the puzzle that composes the united human family. To reach full consciousness there are four laws, and by applying these laws, Earth humans will earn their place in the great galactic plan.

The Law of One   The goal of every being is to discover their soul path for personal growth and service.

The Law of Two   The power of creation can be utilized through loving closeness with another being. This close relationship of caring for another leads to a deeper knowledge of guardianship.

The Law of Three The bonding of one's close relationship with self, friends, family, and clan can develop the web of global or planetary interdependence.

The Law of Four   This law expands the Law of Three to larger groups such as clan to clan, to the planet, and to the solar systems.

Lifestyle

When two Sirians decide to have a child, they will visit a counselor to determine the proper time to have a child. Before conception, they enter into a special meditation and ask the soul who wishes to inhabit the body to be of service. The parents will know why the soul is in that particular service, which will help fulfill the required karmic debt.

Sirians live in a cave-like system of large underground cities built near the power nodes of the planet. Small colonies are found on the surface allowing them to interface with ancient temple sites.

Most Sirian men have a boyish-like figure, and some have bodies of a well developed Earth human male. The males carry a wand-like stick with a glowing ball that serves as a body scanner and diagnostic instrument. Males dress in a full-body jump suit. Sirian females dress in a traditional outfit that indicates what clan she is from. Most women carry a boomerang-shaped object whose main function is to balance the chakras and assist in healing. The Sirian civilization is composed of fully conscious beings.

Their diet consists of a mixture of living vegetables and fruit, with the food being highly energized and easily digested, and meals are eaten twice daily. Prana energy is very high on the planet, allowing them to require only one-half to two hours of sleep daily.

Their life span is between 3,000 and 4,000 years, and when they feel the present life is complete, they ascend into the higher energy patterns. It takes about 70 years to reach young adulthood. During this time they study the Law of One where it is understood and practiced. During this phase of life, sexual relations do not occur. It is a long process to achieve wisdom, taking up to 2,000 years.

For recreation, Sirians go to other star systems or other parts of the planet. They love to dance, talk, laugh, tell jokes, and tell stories. Sirians play musical instruments and compose musical compositions. Those with good voices are constantly singing. Many have hobbies in the fine arts. Sirians love to learn, as it brings great joy and excitement to them. Often they make jokes about their learning passion and even sing about it. Libraries are found in Sirius, but most Sirians have their own immense library of consciousness closely equivalent to the U.S. Library of Congress.

Communication is through the use of visualization and telepathic powers. Sirians have the ability to process information through their mind's eye, and another telepathic being could receive the image and any knowledge associated with it in an encapsulated version.

The Sirians have a clan system for the purpose of enabling an individual to fully achieve his or her life task and inner growth. The clan system has six divisions:

1. The spiritual warrior clan
2. The science clan
3. The science engineering clan
4. The administration clan
5. The life science clan
6. The life science engineering clan

Other star systems follow this pattern, with the Pleiadians having 12 clans, as do the Andromedans. The clan defines the basic service and task that an individual chooses for its incoming incarnation. For example, a person who is of the Life Science clan has the task to understand and maintain the plant and animal life on any world. Their purpose is not to intervene, but to aid and support all life on the planet and solar system. The clan's purpose is to see that the goals and potential of each individual within a specialized interest group are achieved. Each clan has millions of individuals in it with numerous subgroups that may contain between 50 and 60 people, which is experienced as an extended family.

THE GALACTIC FEDERATION

Our universe consists of both light and dark forces. The dark force is spread across the galaxy and has conquered thousands of star systems. Something needed to be done to protect these vulnerable civilizations. About 4.5 million years ago, the Galactic Federation was formed to prevent interdimensional dark forces from domination and exploiting the galaxy. It served as a type of United Nations whose only purpose was to allow light to flow into the Milky Way galaxy. Presently, there are nearly 200,000 star systems and star leagues in the Federation. The primary energy of the Federation is the creation of love. It also provides a defense system against sudden and unwanted attacks. The Galactic Federation also explores, exchanges technology, and has cultural interaction.

There are 14 Regional Councils comprised of planets, star systems, and star leagues. It is an organization of fully conscious and peaceful civilizations. When a civilization has achieved a certain level of technological and cultural development, the civilization is contacted following a scientific evaluation. The Earth has been under karmic laws established by the Pleiadian control of our solar system since the fall of Atlantis 10,000 years ago. The karmic law was overturned, and Earth was granted full membership into the Federation on March 5, 1997. This now allowed the Galactic Federation to establish rescue teams and to formally empower the First Contact team to prepare for contact.

The dark force members comprise the Reptillian Alliance (including renegade Grays and Dracos Reptilians) that has caused chaos in many parts of the galaxy, including Earth. There are former members of the Reptillian Alliance who have seen the light and have asked for an end to all hostility, and some have even requested membership into the Galactic Federation. Over the years, the Reptillian Alliance has attempted to sabotage or attack many Galactic Federation science missions.

A division of the Federation involves liaison groups which have 2 billion workers who act as a primary information network to help members make insightful decisions. The highest level of the Galactic Federation is the Federation Council located in the Lyran star group on Vega. The second level is comprised of the 14 Regional Councils, including the Sirian Regional Council, which Earth will eventually join. Regional Councils establish policy for particular regions and serve as a court of last resort to negotiate misunderstandings.

Lyran-Sirian Elder

There are also local or star system governing councils composed of two types – a local star system council and a star league governing council consisting of 20 or more combined star systems. The largest in the Sirian Regional Council is the Pleiadian Star League that consists of 50 different systems.

The Galactic Federation has been helping to raise consciousness on Earth to save our planet. In the 1960s, the Federation had a difficult time making contact with surface governments and the United Nations. They needed to warn us that we were destroying our planet by our disregard for the environment. They also put in place a First Contact team that would hopefully make contact with Earth with the goal of assisting us into full consciousness. However, they had to change their plan to one that involves a mass ascension. During the late 1980s, the plan allowed the Sirians to alter the sun's polarities to research a method for the first emergence of the ascension process. This then became the primary mission of the Galactic Federation, to oversee Earth's change from the Third Dimension to the Fifth Dimension. Part of the plan may include mass landings at some point in time.

Chapter Four

THE REPTILIANS
Friends and Foes

Reptilian ancestors may be a hard concept to accept, but a large arsenal of evidence shows that extraterrestrial Reptilians may have had a major role in establishing civilizations on our planet Earth. It will be shown that dragons and serpents were instrumental in seeding the civilizations of India, China, Mesopotamia, South America, Central America, and parts of Europe. How can this be when humans look so much different from any reptilian? Ancient civilizations around the world passed down legends about serpents and dragons and their roles in creation. As in most myths, there is an element of historical truth that underlies them. Some of these Reptilians have been worshiped as gods, some were feared, and some were loved.

According to author Jason Bishop, an extraterrestrial investigator, the Reptilians average six to seven feet in height. Like the reptiles on Earth, they are cold blooded, meaning their temperature and metabolism depends on the temperature of the environment. Bishop claims the Reptilians are well suited to space travel because of their ability to hibernate. He reports they have no scales on their body and no sweat glands. They have a short stubby muzzle with catlike eyes.

Earlier, we mentioned there is an element within the Reptilian race that is trying to control humanity and prevent its evolution for its own selfish gain. These Reptilians are called the Dracos and originate from Orion. It will be shown later how these renegade Reptilians are able to manipulate society through fear, wars, lies, and even through our own DNA. Only a few powerful people control major events on this planet, and it is these people that the Reptilians control. Many other

57

Reptilian extraterrestrials are highly evolved and are trying to help our planet in its evolution. However, not every Reptilian has this higher consciousness, which is the case with the Dracos Reptilians. Often it is through conflict that we grow spiritually, and it appears that the renegade Reptilians are giving us that opportunity. Before we discuss current events, let us examine what has transpired in past ages.

SERPENTS AND DRAGONS

Serpents and dragons abound in ancient folklore, much more than any mammal regarding our creation. R.A. Boulay has probably done the best job of compiling the legends and myths of ancient civilizations regarding our Reptilian past in his book titled *Flying Serpents and Dragons: The Story of Mankind's Past*. As he shows, one does not have to look far in the *Bible* passages of Genesis to find clues about our saurian past.

The Bible

Genesis introduces the serpent living in the Garden of Eden long before humans were created. Serpents did all the necessary work to maintain the garden. The Genesis snake was able to converse with Eve and revealed the truth about the Tree of Knowledge. This snake also had intelligence: "The serpent was the shrewdest of all the world's beasts that God had made." According to Jewish legends, the serpent of Eden looked like a man and talked like him as well. The *Haggadah,* a Jewish book summarizing oral tradition, describes the serpent of Eden as an upright creature that stood on two feet and equaled a camel's height. The serpent was lord of all beasts of Eden.

When Adam and Eve were expelled from Eden, Genesis states they wore "shirts of skin." In Jewish legend, the clothes were made of reptile skins that protected them from predators. Serpent skins were symbols of the ruling race, and "When they wore the coats, Adam and Eve were told all creatures on Earth would fear them." The *Haggadah* reveals that the clothes were made of skin sloughed off by the serpent, and that the bodies of Adam and Eve "had been overlaid with a horny skin." Regarding

Adam, the *Haggadah* stated, "It was as bright as daylight and covered his body like a luminous garment. His skin was scaly and shiny and had no need for clothing." When Adam and Eve partook of the forbidden fruit, the major change was the loss of their smooth and shiny reptilian hides and the necessity of wearing clothes. After Adam was created, the *Haggadah* said his pale skin was green.

As long as Adam and Eve remained in the Garden of Eden, they did not propagate. However, once they developed the ability to procreate, this was the "Fall of Man." They took the traits of mammals, and Eve's punishment was to bear the pain of birth like a mammal.

Sumer

Ancient Mesopotamia (Sumer) is considered to be the root of all civilization on our planet, and historians have a saying that "All roads lead to Sumer." Scholars believe that many of the Biblical passages found in the Old Testament originated from the writings of Sumer. Archeologists have discovered countless artifacts of this ancient civilization in modern-day Iraq, with their history being recorded on clay cuneiform tablets. They have withstood the elements of time and provided us with a window to the past. The writings tell of an extraterrestrial race called the Anunnaki who came to Earth from a planet Nibiru, which has a 3,600 year elliptical orbit around our sun. Researchers have been puzzled about who the Anunnaki were and to what star system they belonged. Zecharia Sitchin, considered one of the best researchers regarding ancient Sumer, has written a series of books called the *Earth Chronicles*. Sitchin has not made an attempt to identify the origin of the Anunnaki, except that they came from the undiscovered planet Nibiru, which is of our solar system. Evidence presented by R. A. Boulay suggests they may be of the Reptilian race. Another chapter will discuss the evidence they may be Sirians.

One of the most cherished artifacts found in the Sumer ruins was the Sumerian King's List that traces their lineage to 240,000 years ago. This is when the Anunnaki came to the planet and built a civilization long before the Great Flood.

The King's List, originated in the third millennium B.C., lists all the various rulers and dynasties of Mesopotamia. Archeologists have confirmed king's names of the Third and Fourth Dynasties that are recorded on the list. In that era it seems to be accurate. The List records the length of reigns from the beginning when kingship descended from heaven and founded cities in the Mesopotamian plains. Kingships lasted for thousands of years, but modern scholars discount this information because no one could live that long. However, we have provided information in other chapters on some extraterrestrial societies where they do live for thousands of years. The Bible also talks about the life span of certain individuals in the Old Testament who lived for hundreds of years. However, historians concur on the fact that civilization on Earth did originate in Sumer.

Following the Great Flood, the Sumerian cities were rebuilt and resettled, with the oldest dating back to 3500 B.C. After the Deluge, the water of the Persian Gulf had risen 150 feet because of the Great Flood and melting ice sheets. Boulay believes the Garden of Eden and some of the original cities of Sumer may lie beneath the adjacent Persian Gulf.

The Anunnaki came to Earth for gold to help save the atmosphere on their planet Nibiru. Eridu was the first city built on Earth, probably giving rise to the name of our planet. From Eridu, the god Enki directed all operations. It appears Enki may have been amphibious because he spent the nights in the water. Larek was the third city assigned the kingship, and this was the space control center. Located in the city of Sippur were the platforms for the space shuttles. Nippur was the city where the god Enlil lived, the brother of Enki. Enlil exercised supreme authority over all the Anunnaki on Earth before the Great Flood. Nippur was where Ezekiel saw the "fiery chariot." Every Sumerian city was dedicated to a major god.

The major cities built ziggurats located in the center of a great court which overlooked the city. Atop the ziggurat was a sacred temple or "holy of holies" reserved for the gods when they were on Earth. It was here they mated with selected humans to produce a race of demigods to serve the kings and act as a buffer to humankind. As a result of their genetic manipulation, man lost most of his reptilian nature, according to Boulay, that is, "his shiny luminous skin and scaly hide." Mammalian characteristics were acquired, such as soft flexible skin, body hair, the need to sweat, and the ability to produce live young. Man now had to wear clothes. Perhaps this could explain the Biblical and *Haggadah* writings.

The most significant Sumerian term used to describe a flying serpent with flaming breath was U-SHUM-GAL, a common epithet for the god Enki. Modern translation of this term is dragon. Another Sumerian tablet dated 3500 B.C. is explicit about the reptilian nature of the gods. It reads:

'The reptiles verily descended
The Earth is resplendent as a
well-water garden
At that time Enki and Enlil
had not appeared . . ."

In the Sumerian dictionary of the University of Pennsylvania, the term SHUM means fiery serpent. It may also represent a sky vehicle with flaming exhaust. Boulay concludes that the flying serpents we are dealing with are actually space vehicles of some kind. The Sumerians seem to associate flying with flames. The composite word U-SHUM-GAL means a fiery flying serpent. During the early days of Sumer, U-SHUM-GAL was a complimentary name for a serpent god.

MUSH was another name for the serpent sky gods. Often HUSH, meaning flaming, was added to the term MUSH. MUSH-HUSH applied to the gods who lived in the heavens and we presume to be orbiting in space-craft.

In another chapter, we will discuss in more detail the Anunnaki race and their influence on our planet. Boulay believes the Anunnaki were Reptilians. There is also evidence they may have been from Sirius A. Like Orion, Reptilians probably have other colonies around the galaxy.

India

In the 1920s, archeologists made a hallmark discovery in the Indus Valley of India. They discovered ruins of two large ancient cities, similar to cities of Mesopotamia, that seemed to have been fully planned before they were built. The ancient cities, named Mohenjo-Daro and Harappa, were built between 3500 and 3000 B.C. and came to a violent end around 2000 B.C. Skeletons found in the ruins emitted high levels of radioactivity, as if a nuclear incident had destroyed the city. These people were not related to the Aryans who arrived in India 500 years later. Historians now believe the ruins were home of an ancient serpent people called the Dravidians. One of the oldest Sanskrit sources, the *Book of Dzyan*, tells of a serpent race that descended from the skies and taught mankind. They were the Sarpa or great Dragon people that were the Fifth Race to inhabit the world. The Fourth Race, according to the book, was a race of giants that was wiped out before the Deluge.

The serpent race arrived after the Flood and instructed man in the art of civilization. The serpent gods possessed a human face and a dragon-like tail. Founders of divine dynasties on Earth, they are thought to be ancestors of our current civilization. The leader was called "The Great Dragon." The Dravidians, who lived underground in caves, were also called the Nagas and were said to be a combination of the genetic lineage of humankind and the serpent gods. Described as dark skinned with a flat nose, the Nagas were known for their wisdom. Sanskrit sources describe them as intelligent reptiles that could fly in the sky with their chariots. Some of the great epics of India concern the Aryan contacts with the serpent people, finding some friendly and some hostile. In the epic *Mahabharata,* a group of celestials, the Nagas, attended a wedding feast of an Aryan king. The Nagas and Aryans intermarried, producing kings and heros.

Ceylon, or today's Sri Lanka, was the stronghold of the serpent people, being the island kingdom of the serpent god Ravan. Even in ancient Chinese legend, Ceylon was home of the Nagas. Before the arrival of the Aryan people into India, Ceylon was described as a land of strange reptilian-like creatures.

China

Legends abound in China about the influence of dragons on early civilization. Dragons are often described as benevolent creatures that were friendly to humankind. There were four basic kinds: 1. The celestial dragon lived in the heavens and maintained cosmic order and prevented catastrophes on Earth. 2. Subterranean dragons were in charge of all the precious jewels and metals buried in the earth. Each dragon wore an enormous jewel under the chin symbolizing a pearl of wisdom. 3. The earth and river dragon regulated rivers and prevented serious flooding. 4. The spiritual or weather dragon controlled the weather.

Chinese legend tells of the first humans being created by an ancient goddess named Nukua, who was part dragon and part mortal. Chinese history claims that Asian dragons were present at creation and shared the world with humankind. Dragons taught man how to make fire, to make nets for fishing, and how to create music. Chinese lore teaches that the celestial dragon was the father of the Divine Emperor of the First Dynasty. The Emperor contained dragon blood, sat on a dragon throne, rode on a dragon boat, and slept on a dragon bed.

The most ancient of Chinese books, the *Yih King*, described early days when man and dragon lived together peacefully and intermarried. Many ancient emperors were described as having dragon-like features, like Hwanti, who ruled about 2697 B.C.

Many sinologists believe that Chinese culture originated in Mesopotamia and was a colony of Sumer, along with those colonies established in India, Egypt, and Mesopotamia.

Central America

The snake is one of the most revered symbols among the Mayan people, with most serpents depicted with feathers that symbolize their ability to fly. The Quiche' Indians' legends relate that the first inhabitants of the Yucatan were called Chanes or "people of the serpent," who came from the east and were led by Itzamma. Mayan expert, J. Eric Thompson, claims that the term "itzem," from which the god's name is derived, should be translated as "lizard" or "reptile." The sacred city of the god Itzamma was Itzamac, which literally means the "place of the lizard." The god Itzamma is often depicted as half human and half serpent. He was the most important deity of the Mayan pantheon, who ruled the heavens and was considered the creator god that breathed life into man.

The Aztecs also include a benevolent feathered serpent god in their pantheon, Quetzalcoatl, who is the plumed serpent god that brought great benefits to the Mexican civilization. The name comes from *quetzal* meaning bird with long green tail feathers and from *coatl* meaning serpent. Quetzalcoatl is the most significant Aztec god who taught man about the arts and sciences.

One can only conclude that world-wide depiction of flying reptiles and their legends provide strong indication that possibly our creation and ancestors were of an alien reptilian breed. Nearly all the ancient cultures carry legends of their origin being from reptilian-like creatures. Serpents and dragons, not mammals, dominate these ancient legends and myths.

Underground Dragons

Cultures around the world teach about legends of underground cities and of hidden gates which give the gods access to the underworld. Many cultures trace their ancestors to people and gods who emerged from below the ground following a major world catastrophe. The surface had become too dangerous for them and they had to go underground. The Hopi Indians claim that the Anasazi, their ancestors, emerged from the underworld to repopulate a devastated world.

The mountains of India and Tibet hand down legends about underground tunnels and cities that were created by the serpent gods. According to Hindu legend, there are extensive caves and underground tunnels in the Himalayas attributed to the legendary Nagas, the serpent people. These underground cities were lit by magical stones and were the location of fabulous treasures.

Legends talk about a place called Naggar in the western Himalayas, a city where the kingdom of the Nagas was located. The Hindu classic, *Vishnu-Purana*, tells of Patala, the capital city of the Nagas where the snake gods lived with their brilliant jewels, and where a great library existed regarding antediluvian days.

Theosophist leader Helen Blavatsky spent three years in Tibet compiling writings of Sanskrit from a book entitled the *Book of the Dzyan*. She reports stories about subterranean caverns with concealed entrances containing artifacts of the sky gods. There were secret chambers of all kinds connected by miles of tunnels. She describes the Nagas as semidivine beings who had a human face and tail of a dragon. The *Book of Dzyan* reports about an ancient race of serpent men who descended to Earth and founded the Fifth Race on Earth, giving birth to the present civilization. The ancient Sanskrit literature also reveals a struggle between the Nagas and ancient people. It tells of an underground city called Bhagavata, home to the serpent people, without sun, moon, or stars but lighted with sun and moon stones.

South America is laced with miles of tunnels reaching from Cuzco to Tiahuanaco, to Machu Picchu, to the Pacific Ocean. Cuzco is the ancient city of Peru founded by a serpent god. The tunnel system is said to have been built by the people who built the megalith stonework of Machu Picchu. The Indians talk about a system of caves and tunnels in Ecuador that extend for hundreds of miles. Metal and stone tablets with unknown hieroglyphics have been found in the caves.

A REPTILIAN ENCOUNTER

Stefan Denaerde's life changed forever one day in the 1960s when he was sailing with his family in waters off the Netherlands. Stefan, a successful businessman in the Netherlands, wrote a book about his experience entitled *UFO – Contact from Planet Iarga* regarding his encounter with a group of Reptilians. He used a pseudonym because he would have been recognized in his country, and for business concerns he wanted to remain anonymous. Wendell Stevens, America's top UFO researcher, investigated the case for four years and concluded the facts to be true and coauthored the English version published in 1982.

Confirming this encounter was NASA, which had picked up an incoming radio frequency with an unusual bandwidth near the Hague, confirming a possible contact. The Iargans, the Reptilians who made contact with Stefan, had contacted four other Earth humans in a similar fashion and imparted similar information.

Stefan's experience began when he was sailing with his family on windless waters of the Oosterscheht in the Netherlands. The boat's compass suddenly malfunctioned and out of the darkness, a blue-white searchlight shone directly into his eyes, originating in front of the bow. About the same time a high-pitched whining noise was heard when the yacht came to a standstill against something solid in three feet of water. Stefan shined a light into the water and saw the hull of an overturned ship and a body floating face down. Grabbing a rope and dinghy, Stefan jumped into the water to rescue the body. He then noticed a diffuse light under the water's surface and a dark shape wading quickly toward him, similar to the body he went to fish out of the water. Becoming fearful, Stefan panicked and rushed back to the yacht. He noticed the being pulling the dinghy with the body onto a platform and lifting the body in his arms. Back on the yacht, Stefan again shined a flashlight into the water and saw a green thing about 50 feet in length that had become magnetically attached to his yacht. From the submerged craft, two beings crawled out of a hole and greeted Stefan in a friendly manner. Stefan described the beings as five feet tall with short legs and long arms, both wearing a smooth, seamless metallic suit. They asked Stefan if he could speak English and thanked him for rescuing the crew member. The beings told Stefan they were from another solar system and had been studying the planet Earth. As a token of thanks, they gave Stefan a block of inert metal, only half as heavy as Earth's best steel. It had superconductive properties allowing a current to flow between a positive pole and opposite negative pole. They said the metal's melting point was higher than anything on Earth. It was the same metal used on the outer skin of their spacecraft. They wanted to prove to Stefan that they were for real.

The aliens said they had not made contact with Earth because humans do not know the laws of a higher civilization. They told Stefan that the most important natural law in a highly technological society is to eliminate all discrimination, as discrimination blocks cosmic integration. Cosmic isolation of an intelligent race can be lifted only when a minimum social stability has been reached. The beings then gave Stefan a choice of keeping the metallic object or two days of knowledge on board the craft. After talking to his family, Stefan chose the latter. They told Stefan that on board the space-craft, the aliens had a small sterilized decompression pressure chamber from where he would be able to hear and see a screen. They warned Stefan that after two days, he would become wiser but not happier. Stefan had to take a vow that he would not contact anyone. Stefan and his family agreed, and the yacht was anchored.

The space-craft emerged from the water and Stefan was ushered aboard. He was told the language they speak is the language of all living species in the universe. It was spoken on Earth before the Babylonians confused the tongue. They told Stefan, "You don't hear words, but sounds that are directly reflected by your emotional structure, the life field. Therefore, don't try to understand words, but listen to the reflection of your soul." They asked Stefan if he would like to see them face-to-face. Stefan agreed, but became quite fearful when they revealed themselves. He saw eight humanoid beings with faces looking like a primitive animal. The head was about the same size as ours, and a bony ridge ran through the middle of the skull that had a deep groove in the center of the forehead. The neck was much thicker than ours and their body more solid, with broad chests and short stocky legs. Their hair was short and smooth with colors of rust brown, gold, and silver gray. The skin was hairless and gray-brown in color.

They told Stefan that Iarga is located ten light years away and that they are of the same origin and identity as Earth humans. Their planet is larger than Earth, with a dense atmosphere and heavy rain in which humans could not survive. The planet is almost covered with water with a landmass about the size of Australia. Their population is 6,000 people per square kilometer, 100 times more dense than Earth. Because of the high gravity field, their bodies are short and compact. Originally, they were amphibians who belonged in the water. Their hands and feet are large and broad with webs between their fingers and toes. Their sex drive is much less intense than humans'. They lack full lips, ear lobes, protruding female breasts, and external sex organs.

They told Stefan that only beings who possess the ability to improve their mentality and eliminate aggression have a chance of reaching perfection on such a planet. Their intellectual, emotional, and creative capability are about the same as Earth humans. However, they confided to Stefan, they consider their weak point to be their individuality, because they do everything in groups. They think col-

Iargan

lectively and obey the law of their society to the letter. Iargans live for the love and friendship of the group.

The Iargans wanted to demonstrate to Stefan the secrets of a highly developed culture by means of a holographic film which would show the planet Iarga to Stefan. Freedom of thought is the essence of humanity, they told Stefan. Their purpose was to convey only knowledge, not convictions. Psuedo-conviction paralyzes individual freedom, making humans rigid and dogmatic.

They have an instrument called a radiation reflector, and on a screen they showed an explanation of a topic in the form of a picture story. The device was able to fix Stefan's concentration through its radiation, and Stefan did not have to take notes. The information gained through the radiation device remained locked in his memory forever. The knowledge was fed directly into Stefan's brain. It was a combination of visual stimulation and thought transference.

Lifestyle

The Iargans live by three principles that underlie their high level of culture – freedom, justice, and efficiency. Without efficiency, their world would collapse. Efficiency is almost a religion to Iarga. For example, their efficient farming techniques produce the maximum amount of food. Their justice would fail if a person's house showed a difference in stature from his neighbor's. Everyone on the planet lives in the same type of house and rides on the same trains and same cars. Transportation is mainly by train, through a fully automated robot rail system. The slim torpedo train moves without friction and is the cheapest form of transportation. Iarga is a world without refuse, pollution, odor, exhaust gases, or traffic jams. It is also a world without money exchange. All the people are taken care of. They emphasize that culture is the measure through which a society caters to the least fortunate. The measure of collective unselfishness is what makes an intelligent race immortal – how one takes care of the sick, invalids, old, and poor.

Their domestic lifestyle is reflected in their efficiency. Housing complexes are in the shape of a ring and designed to last 1,000 years. They use the planet's heat for their power supply. Most Iargans work at home, which prevents unnecessary transportation. Fisheries are a major food industry in Iarga, with species of fish similar to those we have on Earth. They also eat meat. They do not consider the planet overpopulated if everyone has enough to eat. Both men and women work as equals around the house, and chores are shared by everyone. Shopping is done by a computer.

Mindset and Health

Iargans are constantly trying to make their world a better place to live. A race that lives for the future is concerned with the highest efficient use of natural resources. Races that live under constant threat of war do not make logical plans for the distant future. They think Earth humans live too much for the present and hang onto the past.

One of the greatest problems on Earth is discrimination against others, according to the Iargans, who said humans seem to be continually occupied with thinking of new forms of separation. They emphasize that it would be possible to change so many things in our world if we just stop discriminating against others.

As most advanced civilizations, the Iargans believe there is a mind-body connection and the mind can cause health problems. Each patient in a hospital is connected to a computer that caters to individual needs, such as pain reduction, medications, entertainment, etc. The Iargan philosophy is to prolong happiness, not life when the end is near.

Economy

Stefan was told by the Iargans that their economy "is aimed at efficiently satisfying a man's needs so that he or she is released from the tyranny of procuring material things over his daily life." This is achieved by providing equal shares for everyone and the culture becomes quite stable. They can achieve this through two possible methods: 1. Everyone must own the same amount, or 2. No one can own anything. The latter they found more efficient. Because money is a form of property, it has been abolished. Personal property indicates a primitive level of culture to the Iargans.

A huge number of companies, called trusts, control the total production of goods and services. These large organizations each have millions of employees who are active over the entire planet. Primary trusts distribute directly to the consumer, and secondary trusts supply the primary trusts. Nothing is paid for on Iarga, only registered. Nothing can be purchased, only hired, through what is known as the right of acquisition, including housing, boats, and cars. Following death, goods are returned to the trusts. Legally, everything belongs to the trust, which is responsible for upkeep, repair, and warranty.

The goal of their universal economic system is to create a natural leveling off of income. During the early stage of social stability, this is not possible and a material incentive must be offered to stimulate a greater personal effort. A social minimum must be determined in the beginning, while security for the young and old must be established. The whole purpose of their economy is to free people from material motivation. Everybody on Iarga is rich, and there is no upper or lower class.

Their emphasis is to free the individual as much as possible from noncreative service. When production reaches overproduction, the work day is shortened to allow creative pursuits.

By developing their plan of great efficiency, Iargans reach the point where everyone works only one day in the week on the direct production process. By volunteering to restrain their consumption, equality of noncreative work output has led automatically to the equalization of incomes. By waiving their right to consume, their needs decrease. Then comes a moment in their development when controls of consumption are lifted and all goods and services are available to all above a certain age. Individual self-discipline has come of age and material greed is conquered. Having free access to all the prosperity for everyone makes it impossible for an individual to be wanting.

Family and Friends
Iarga is a planet where people love each other and are happy to meet each other socially. There is a lot of hugging among spouses and children. All the Iargans have the same duty to the children in the group in which they live. In the schools, knowledge is planted in the children by the radiation device just discussed, but the adults must help the children transform this knowledge into experience.

As soon as children reach sexual maturity, the parents arrange for the child to undergo a psychological and medical test. If this is passed, the child is declared legally free and can vote and have sexual freedom. Iargans do not have the same pleasure in sex as Earth humans, but value more the experience of intimacy and love. There is precise population control on their planet.

Men and women are considered equals on Iarga, but with different mandates. Women maintain a dominant position because they must lead the mental development of the children. They demand respect as individuals with intellectual sovereignty. The relationship between sexes is directed at creative expression for one another with sex playing a minor role, which does not affect the relationship.

Spirituality
Iargans believe they are bound by the Law of Cause and Effect and are subject to reincarnation principles. Reincarnation selection exists on Iarga, which roots out the adepts of evil. They

believe that large doses of unselfishness can exist only in an environment protected from evil, so they have designed their environment just to prevent evil.

Iargans speak of the threefold purpose for existence:
1. The creation of individual identity.
2. The creation of their immortality by use of their talents and attempting to reach self-chosen creative goals.
3. The development of their secondary identity which is the culmination of their daily choice between selfish and unselfish creativity.

They believe a superculture can be recognized mainly by its unbridled creative power. Happiness and satisfaction mean reaching the goal of one's creativity.

They told Stefan that the Iargan race, which includes a population of billions, differs little from the human race on Earth. However, they know only one goal, which is the perfection of their society by mutual love. The Iargan race has reached a high level of love, knowledge, and wisdom.

Space Mission

The Iargans told Stefan that a space mission lasts about 20 years. Their space craft is fueled by water. Iargans obey the law of interplanetary contact of noninterference. The free will of a cosmic race may never be infringed upon. Iargans are allowed to plant knowledge, but not to exert any pressure to do something with the imparted knowledge. This knowledge is planted in the collective consciousness and eventually will come to the surface. Human free will is honored. Knowledge can influence the freedom of choice of an ignorant race, especially when presented with authority. This is why Stefan was told not to prove the existence of Iarga. However, he was instructed to publish this knowledge in a number of languages. He was never to convince people of its truth. After two days on the craft, Stefan departed and rejoined his family on the yacht. Because of the agreement with the Iargan's, he took the film out of his camera and destroyed it. Stefan and his family watched as the spacecraft left the water, and Stefan knew his life would never be the same.

THE DRACON REPTILIANS

Not all Reptilians are of the Light, and one race that has created chaos on Earth for eons has been the Dracons from Orion in the star system Sigma Draconi. The Dracon race is a sentient, intelligent race of upright standing beings with a dragon/lizard appearance and a tendency toward aggression and warlike behavior. The Dracons have caused the death and destruction of

many planets and were created by the Creator to represent the masculine and dark side. They lack spiritual development and are a digression from a superior Reptilian race. This group has created a very disruptive genetic imprint on Earth's human DNA. When the Dracons visited Earth during periods of the first, second, and third seedings, they created strains of their own species that could live comfortably in Earth's environment. The hybrid

Dracos (Drakon and Human Hybrid)

root race that they created is called the Dracos, who have very aggressive tendencies. The Dracos hybrids, whose body structure resembles humans, were more lizard-like in appearance, but with facial and temperament characteristics that resemble the Dracons.

Ancient History

During Earth visitations, the Dracons began to mutate genes from captured humans by seeding the women through painful inductions. Although many of these hybrids had difficulty surviving within the Earth environment, some did. These hybrids became known as the Dracos Reptilian race. The Dracos began interfering with the frequency fence that protected the Host Matrix Soul codes, as discussed in another chapter.

The Dracons also tampered genetically with certain species of dinosaurs that had been seeded by other extraterrestrials as an experiment. The hybrid dinosaurs were monitors who were aggressive and carnivorous. Many of the cultures of the second seeding had to escape the terror of the Dracon monitor dinosaurs. Realizing what the Dracos were doing to humanity's evolution, the Guardian forces banned the Dracos from Earth, temporarily removing the Dracon problem of interference.

As members of the human lineage, they felt entitled to use the territories of Earth, even though they were banned nearly 850,000 years ago. The intrusive visitors of the Dracos lineage are in reality a hybrid mutation of our ancestors who evolved within the Orion star system. The soul essence of the Dracos is part human and part Dracon.

The Dracon and their creations had become a serious problem to the developing races, so plans were developed to dispose of the Dracon menace. Approximately 950,000 years ago with assistance from the Anunnaki visitors from Sirius A, extraterrestrial races formed a coalition to use the power within the Earth's grid to destroy the underground habitat of the Dracons and their monitors. The plan backfired, and the explosion resulted in a small ice age. Because of the ice age, the Dracons left Earth during this period. Most humans retreated underground and created sophisticated cultures beneath the Earth's surface, and some races were permitted to enter the portals of Inner Earth. Even though most of the Dracons had left or perished, they left a legacy of genetic distortion on Earth that threatened the continuation of the human lineage.

About 52,000 years ago, the Templar Annu allowed the Dracos to secretly return to Earth. The Dracos infiltrated the Lemurian continent of Muarivhi and created an extensive network of tunnels between Lemuria and Atlantis and proceeded to terrorize both continents. The humans hoped to use the generator crystals to seal the Draco tunnels, but the explosion destroyed the land mass of Lemuria and its civilization.

Current History

In 1926, time traveling members of the Zeta/Zephilium races, who were originally from Apaxein-Lau (Apex planet mentioned earlier) in Orion, began interacting with private factions of several Earth governments. Their diabolical motives created a very negative effect on the vibration of the Earth. If the Sirian Council had not intervened, Earth populations would have been decimated between the years of 1972 and 1979, and a pole shift would have occurred in the mid-1980s. The primary goal of the Zeta and the Dracos (Dracon-human hybrid) and Rutila (Zeta-Dracos hybrids) was to take over Earth's territory. The Dracos believed that dominion of Earth is their birthright. They were Earthlings who had been banned from their home planet 850,000 years ago and have existed on Orion since.

Several Zeta and Dracos groups have refused to give up the goal of controlling Earth, and others formed a coalition called the Dracos-Zeta Resistance. The hybrids between these two races are known as the Rutila, those beings referred to by the U. S. government as Extraterrestrial Biological Entities (EBE). They look much like the grays but have a lighter gray or gray-white complexion.

The Dracos-Zeta Resistance had to find a method to construct the Zeta Seal, a frequency fence, and Zeta mind complex before 2012 in order to regain control of the human population in Dimension Four. In 1983, they set their plan in motion by motivating humans to create another experiment similar to the Philadelphia Experiment of 1943, when the U.S. government was trying to make warships disappear visibly, but accidently sent some of the crew into the future and ripped the space/time fabric. Upon the return of the U.S.S. Eldridge from the invisibility experiment, some of the crew returned with their bodies enmeshed in the bulkhead, causing the government to cancel the research.

By creating another rip in the space/time fabric, the Resistance would be able to send space-craft in from the future. These ships would enter the Dimension Two frequency band and transmit electromagnetic pulses to reconstruct the frequency fence and cause mutations in the Fourth DNA strand. This would prevent the human gene code from assembling the Fifth DNA strand and prevent the scheduled ascension through the Halls of Amenti. The Dracos-Zeta Resistance would have to begin broadcasting no later than 2006. Working with the Secret Government, the

Dracos-Zeta Resistance orchestrated another experiment called the Montauk Project located at the tip of Long Island in New York. The time period of 1943 (Philadelphia Experiment) and 1983 (Montauk Project) were successfully linked to the Dracos-Zeta Resistance Dimension Four time period.

From 1983 to the present time, the Dracos-Zeta Resistance has resumed the hybridization program by abducting humans and creating several strains of hybrids and human clones. Through genetic engineering, they have created infiltrates through which they could interfere with Earth cultures in the guise of human form. The infiltrates are children that were conceived by natural conception. The mother was abducted during pregnancy and the Zeta-Dracos genetic material was infused into the fetus. These children are born after a seven-month gestation period, raised by their Earth parents, and appear fully human. They are consciously aware of their ET affiliation, and the Draco-Zeta Resistance can subliminally direct them.

In 1984, the Guardians found out about the Montauk Project and its potential destruction. Guardian groups joined together to create the Bridge Zone Project, which involved shifting Earth completely out of Harmonic Universe One time cycle, out of the way of the Dracos-Zeta Resistance. Once Earth was stabilized in the Bridge Zone 3.5 time continuum under full Guardian protection, the Dracos-Zeta Resistance could not successfully launch an infiltration project, losing control of the frequency fence and human capture of Dimension Four. To move into the Bridge Zone, the population would have to assemble their DNA to the 4.5 level and a minimum number of people would have to assemble the Fifth DNA strand. There would also need to be 144,000 individuals to fully assemble the Sixth DNA strand, embody the entire Soul Matrix, and then begin assembling strands seven through twelve. This would release the Dimension Four Zeta Seal and allow the Earth's grid speed to rise enough to remain in the Bridge Zone. If the Bridge Zone Project is successful, humans will be free from control of the Draco-Zeta Resistance.

If the plan fails and we fall under Dracos-Zeta control, our souls will then devolve on a Descending Planet. If the

Secret Government refuses to cooperate with the Dracos-Zeta Resistance, they will be unable to orchestrate a full infiltration into human society.

REPTILIAN CONTROL

Evidence suggests that only a handful of people control major events on Earth, and this is through the Secret Government that controls the economy, wars, and politics. Most of these individuals are the super rich who have formed a coalition among themselves under various names that will be discussed later. These coalitions have resulted in the Secret Government that controls at some level the super governments around the world, including the United States. Some members of the Secret Government are in contact with negative extraterrestrials, who have their own agenda. They have received advanced technology in exchange for favors given to the extraterrestrials. Other members are controlled through their corrupted DNA. Members of this Secret Government are composed of mainly the Illuminati, who contain this corrupted DNA. Politicians with the corrupted DNA have been controlled by the extraterrestrials and have been placed in office for that particular reason. David Icke writes about Reptilian Control in his book titled *Alice in Wonderland and the World Trade Center Disaster*. He claims that events of 9/11 were the results of an inside job with government involvement under Reptilian influence. Icke believes that the Reptilians are in control of the world and explains how. The Dracos are back!

According to Icke, the Reptilians "operate just beyond the frequency range of the five senses and use apparently human physical bodies to manipulate the five sense world." He claims "the bloodlines placed in position of power throughout the world are not human as we understand them," citing a number of witnesses who have seen human transformation into a Reptilian body, similar to shamans shape-shifting into animals. Between the Third and Fourth Dimension are crevices of frequency where Reptilians and other entities reside in this interspace plane. Ancient writings found in the *Emerald Tablets of Thoth* confirm this possibility, saying that in the spaces are serpent beings who manipulate the world. The Zulus of South Africa believe similar legends of Reptilian beings that reside in this hidden dimension and manipulate the world.

Witnesses who have experienced Illuminati blood and sacrificial rituals have told Icke they have seen humans shape shift into Reptilians. Rituals are very important to the Reptilians, as they allow the Illuminati to connect to their masters (the Reptilians) in these interspace planes. The human blood makeup (DNA) and vibrational fields created by the rituals produce a frequency environment where the Reptilians and other entities can manifest in the Third Dimension.

Bloodlines

Millennia ago, extraterrestrials interbred with humans to create hybrid bloodlines. The *Bible* says the sons of God who interbred with the daughters of men created a hybrid race called the Nefilim. Wherever the Nefilim went, they created chaos.

Inside sources have told Icke that the world is controlled by bloodlines that are Reptilian, not human. These human/Reptilian bloodlines are called the dragon bloodlines. Chinese emperors acknowledge that their right to the throne was their genetic connection to the serpent gods. Even today, royal families around the world are in power because of their DNA bloodline. Currently, this special bloodline dominates politics, banking, business, and media ownership. The interbreeding continues, and this bloodline is found in the people who control the world, the Secret Government.

People with this bloodline do not display the same emotional response as the rest of Earth humans. These individuals can bring harm to any number of people without feeling emotional baggage. An example is the killing of thousands in the World Trade Center that led to the wars in Afghanistan and Iraq, in which a Johns Hopkins October 2006 study said that over 600,000 Iraqi civilians have been killed. We are told by those influenced by the Secret Government that around 50,000 civilians have been killed. The corrupted DNA of these elite are in a vibrational resonance with the Reptilian entities found in the interspace plane. This is why so many of the U.S. presidents are genetically connected, especially with royal families of Europe. Illuminati families are obsessed with interbreeding to protect the corrupted DNA.

Fear

David Icke says the Reptilians operating in interspace create an energy source for themselves called fear, which creates a vibration that they thrive on. The more humans feel fear, the more energized the Reptilians become and the more power they feel. Scientists acknowledge the most ancient part of the human brain is the part called the Reptilian brain, which represents the core of the nervous system. It is through the Reptilian portion of the human brain that the Illuminati can control humans through fear. Violence, sex, racism, and hatred come from the Reptilian part of the brain. Icke claims that Bush senior, Kissinger, and Cheney are Illuminati, as they have helped create a fearful society where people are focused on physical and financial survival with a Reptilian instinct. The desire for excess derives from the Reptilian brain.

Ritual

The Illuminati, who control the Secret Government, are obsessed with ritual, and Reptilians communicate through imagery and symbols, similar to the Illuminati. Their secret language is based on symbol. The televison and movie industries are owned and controlled by the Illuminati, claims Icke. They create visual images that can control and condition populations.

Activating the Bloodline

According to David Icke, the major Illuminati families know who they are and their hybrid nature. Once the DNA comes into contact with a central vibrational code, it becomes activated and these individuals can be possessed by interspace Reptilians. The corrupted DNA lies dormant until it is activated. However it has no effect on the person when dormant.

To activate the code, one only has to attend a secret society ritual of the Illuminati network. These rituals are designed to activate the corrupted DNA and allow interspace entities to possess. Satanism is one arm of the Illuminati, and they allow their bodies to be possessed during rituals.

The Illuminati have detailed records of individuals who have dormant DNA corruption. One such record is maintained in the genealogy library of the Mormon Church in Salt Lake City, which is controlled by the Illuminati. They also use another method to monitor global DNA under the guise of stopping crime or terror around the world. Law enforcement agencies maintain records of DNA that can be used by the Illuminati.

Once the Illuminati identify individuals with the corrupted DNA, they can influence their careers in politics, military, business, and media. Often the individual is asked to join a secret society to further his or her career. Rituals of these societies can activate the DNA. One such society is the Freemasons with about 99 percent or more of Freemasons having no idea what the initiation rites mean. As the DNA codes are activated during the ritual, a person can be possessed by a Reptilian, and their emotions, attitude, and thoughts begin to change. Many former U.S. Presidents have been Freemasons. Another secret society is the Skull and Bones Society of Yale University whose members exert an exceptional amount of world power.

Icke believes that the leaders of our country are not Americans but Reptilian entities possessing them. This is true for other countries as well, with most of the major world leaders being possessed by Reptilians. The Reptilians manipulate the nation to fight senseless wars and advance Reptilian agendas unknown to our possessed leaders. They advance global centralization and create fear to advance their agenda. The Illuminati is the driving force behind the global fascist state. On a spiritual level, Reptilians do not want people to know their real purpose in life or why they are here, as they want to keep them ignorant so they can easily control them.

How do we overcome this hidden agenda by the Illuminati and Reptilians? Icke thinks the only way to dismantle the Illuminati is to think it out of existence by loving them into a higher state of reality, beyond survival and into the infinite.

The Illuminati are fearful of not surviving, and we need to reassure them that we are not a threat. Icke believes we should give love to the Illuminati politicians – Bush, Cheney, Blair, and Kissinger. Some say to give the love to the Higher Selves of these individuals and not empower the Lower Selves. We have created this world, and we can change it any time we want.

A REAL LIFE REPTILIAN ENCOUNTER

Many people after reading the previous section are quite skeptical, I am sure, about Reptilians controlling our lives. James Waldin, Ed.D., a former college professor in Arkansas and Missouri, would disagree with you. His experience with the Reptilians changed his life forever, and he wrote a book entitled *The Ultimate Alien Agenda: The Reengineering of Humankind*. His experience began in March of 1992 when James, age 45, was about to fall asleep and an alien entered his bedroom. The alien was between four and five feet tall with a bubble head and gray skin that appeared cool, moist, and leathery. The alien shot a brilliant beam of red light from his right eye that penetrated Walden's right leg causing a stinging pain. He ran to the bathroom, but a telepathic voice told him to return to bed, which he did, as if drugged. He next saw a beam of light above him and then lost consciousness.

James awakened upon a cold metal table and was being examined by a group of beings. They examined every part of his body and then extracted a sperm sample. A telepathic voice told him he was in an underground facility beneath a wheat field in southeast Kansas, where both aliens and Earth civilians were working side by side. The beings assured him that he would not be harmed as they were only conducting research. Following his return to home, he became an emotional wreck and even considered suicide.

When told a friend about his experience, she said she had also been abducted and had a device implanted into the nape of her neck. She also had experienced emotional distress and had visited an alien abduction researcher named Barbara Bartholic, who had helped her. James then made an appointment with

Barbara who proceeded to hypnotize him. Under hypnosis, he could remember the aliens saying that he was one of them and that he needed to remember who he was. During the second appointment he remembered that they told him he was part of a human embryo experiment and that the aliens had created him. He began as an embryo from an alien laboratory and had been implanted in his mother's womb. He was a hybrid, part alien and part human.

Under hypnosis James remembered being abducted as a child, when the aliens had inserted an organic implant in his ear that functioned as a communication device. James said the aliens could block his thoughts and control his mind by activating this implant. The aliens, he said, were trying to evolve humans to their level of intelligence. James then had thoughts that he was an alien living inside a body. It was later confirmed that he was a Reptilian hybrid.Through many hypnosis sessions, James discovered that he had survived hundreds of alien contacts, and he now wanted to write a book about it.

In 1992 in Arkansas, following the elections, James saw five athletic men jogging with one man in the center, while the others formed a protective square around him. The man in the center who had just been elected to a very high office, looked up and made riveting eye contact with James. James thought inwardly, "I recognize you. You are one of them. You are a human hybrid." He asked himself, "Are aliens programing hybrids to assume control of the world?"

Most hybrids don't have any conscious knowledge about the inter-dimensional creation of alien programming. Each human hybrid receives special programing; in James' case he was to be psychic. James acknowledges that most hybrids are psychic because of communication devices that have been implanted in their bodies. Psychic ability may only be alien technology. He discovered the alien scientists can program personality, intelligence, creativity, and physical features. They can activate a program at a specific time, including becoming President of the United States or falling in love. These alien scientists strategically located hybrids in communities around the world. James believes that anyone could be a hybrid, from political leaders to gang leaders. He also believes that alien scientists may be orchestrating many world events by manipulating human hybrids.

James later discovered that he had an inter-dimensional son whom he met. Barbara told him that hybrids of James' generation had become emotionally crippled by the experience. Barbara believes that some alien human hybrids are programmed into polarized groups. The groups are then manipulated to oppose each other. For example, some of her patients were Nazis in their past life, and some were victimized by the Nazis, and both groups were hybrids. James concluded that the aliens have been programming people to fulfill their master plan for humanity.

Barbara has found in her research that aliens program our emotional responses to produce misery, jealousy, passion, and love. Love obsessions can be crippling. In fact, James believes that anyone that goes through a traumatic love obsession may be a human hybrid. These strong emotions energize the aliens in some way, and they can manipulate or activate these emotions. Barbara has discovered that most of the abductees were sexually molested before puberty. This was true with James who was raped at age six by his babysitter. Barbara's research with abductees concluded that alien beings are responsible for sexual abuse of children, and they are present during rapes. They seem to experience our feelings and thrive on our fearful emotions and passions. She has also documented their presence during birth and death, and concludes that aliens seem to be involved in every aspect of our lives. James goes on to say that alien beings are present with us so often because they are us. He believes we are actually a multiplicity of beings.

James felt he had two bodies, one being human and the other Reptilian. He said, "My permanent body looks like the aliens, and lives in both dimensions. It's my permanent body, and it will live beyond my human lifetime. When my human life ends, my alien body will return to the other dimensions."

Under hypnosis, James talked about his Reptilian race not being very beautiful and believes the hybridization of humans with the Reptilian race is to create a more beautiful species or race to facilitate the adaptation to the Earth's atmosphere. When he saw his Reptilian body, he was shocked because as a Reptilian he was between 8 and 12 feet tall. He had scaley Reptilian feet and bony legs. He said that air passes through a small opening on the side of his neck, much like gills. The feet are slender and about six inches longer than humans. The body is flexible and loose. Because his alien body is so old, having been around for thousands of years, the head is much larger and the back has a large fin-like appendage. The skin is more of a reddish tone, with shades of yellow, green, and orange.

When the Reptilians arrived ages ago and took control of the Earth, they had a difficult time reproducing in the beginning. Because they lived in subterranean shelters, they also were quite sensitive to light. Reptilians have underground devices that serve as satellites and are used as guidance and communications. They have numbered the various Earth grids and, when they as inter-dimensional beings want to travel, they visualize the number of the anchor grid, and in the blink of an eye, are at that location.

He said the Reptilians controlled everything on Earth eons ago. They were the ultimate authority, but their power has been diluted. James said the Reptilians are gradually passing the responsibility to humans. He thought to himself that perhaps he could be receiving disinformation from the Reptilians to further their agenda on Earth because he was going to write a book about his experience. Through Barbara's research with the abductees, she felt that the Reptillians must be running the whole show on Earth. However, James said that because the inter-dimensional Reptilian race is no longer in absolute control, their intelligence is being dispersed throughout the human population. He believes that no group will be able to seize control of the entire planet.

Because of the hybrid program, the human hybrids who share the blood of humans and intelligence of the Reptilians will eventually manage the planet without guidance from the Reptilians. The hybridization process started long ago, and James thinks the program may be approaching its conclusion. James was told by the aliens that it will take 50 to 75 years before humans are capable of achieving world peace and that the highest intelligence on Earth will not prevail unless all races coexist peacefully. James thinks the ultimate goal of Reptilian intelligence is to populate Earth with integrated human hybrids. This means people, who have human minds and bodies, would manifest multidimensional Reptilian intelligence. Within 50 to 75 years, he believes, all humans will share the Reptilian intelligence. It is the human hybrids who are the carriers of this intelligence and their offspring who will disperse it. The human race is being replaced by human hybrids.

Chapter Five

THE ZETAS
See You in My Dreams

The most recognized extraterrestrial group, commonly referred to as the Grays, is from the star system Reticula Rhombolis (housing Zeta Reticuli 1 and Zeta Reticuli 2) located in Orion.

Zeta Gray–Rigelian

Also located in Orion is a star system called Rigel, a double bluish star, where the negative Grays originate, sometimes called the Rigelians. For the purpose of this chapter both groups will be referred to as the Zetas. The media has often portrayed the Zetas in beer and car commercials, an indication that demonstrates that extraterrestrials have become entrenched in the consciousness of the mass media. People who have been abducted most often describe their abductors as the Grays. Bodies of Grays were found in the Roswell UFO crash during the late 1940s. As we have previously alluded, the Zetas are on a mission, and Earth humans are part of that mission.

The Zetas are between three to five feet in height. Their skin is described as various shades of gray, white, and beige. They exhibit no body hair, and a head that is larger in proportion to the body than humans. Some abductees describe the Zetas as having a very small nose, while others report no nose at all. Their eyes are large and dark with no visible pupils. Males cannot outwardly be distinguished from females as there are no external genitalia. Zetas exhibit no emotions and appear neutral with no signs of being either benevolent or malevolent.

Zeta Gray

Previously, we discussed the humanoid race, including the Zetas, emerging from the Lyra star system. The planet Apex (Apaxein Lau) was the ancient origin of the Zetas, a planet that was similar to Earth. The Zeta's technological progress surpassed their spiritual progress, which led to catastrophic events. They destroyed their planet's surface with pollution and a nuclear war that forced the inhabitants underground to escape the radiation, thus becoming independent from the surface ecosystem.

As their civilization evolved underground, the natural birth process became difficult because the cranial size of the infant became too big to pass through the female pelvis, which resulted in many Zeta deaths during childbirth. Their population began decreasing until they solved the problem by reproducing in the laboratory, eliminating the need for the sexual act, conception, and natural birth.

Because so many Apexians (Zetas) were dying from pollution and radiation, they began to alter their genetics so that newborns would become more adaptive to underground conditions. The bodies were restructured so they could absorb frequencies of light beyond the visible spectrum and redeveloped so they could ingest nutrients from some of the luminous underground rocks. Finally, their birth rate began to balance out their death rate, so the race was going to survive.

Following the nuclear holocaust, the radiation began breaking down the planetary energy field on a subatomic level, creating an electromagnetic warp that collapsed the time/space fabric surrounding the planet. This resulted in a shift of the time/space continuum, and the planet moved to a new time/space continuum in the Reticulum star system. Because the Apexians were underground, they were totally unaware of this shift.

After hundreds of years, the Apexians developed into different cultures with positive and negative consequences. Their bodies became shorter to adapt to the underground conditions. Both their reproductive and digestive systems atrophied, and they were no longer taking in solid nutrients by mouth but through their skin. Their pupils mutated to cover the entire eye, allowing them to absorb light below the visible spectrum.

Realizing it was their unchecked emotions that were responsible for the destruction of their planet, they no longer allowed emotions in their life and genetically engineered them out of their makeup. Passion was no longer going to rule and control them. They did not allow diversity in their culture, so societal conflicts were no longer a threat. Neurological structures were redesigned so that each external stimulus produced the same reaction in every person. Dedicated not to repeat the past, they eliminated war by eliminating emotions which had ruled their culture.

However, not everything was rosy in this underground civilization as two groups of Zetas developed. One group was the positive Zetas, who were benign and benevolent, and the other group was negative and wanted power.

Upon returning to the planet's surface, they realized the time and space shift. The positive Zetas wanted to find out how this happened and diligently learned all they could about the folding of space and time. The negative Zetas built space-craft with their past technology and moved to other planets in the Reticulum system, where they developed their culture. Others explored the galaxy and set up colonies in Orion and Sirius. Much of the material from this chapter comes from Lyssa Royal in her book *Visitors From Within* and Anna Hayes'* book *Voyagers: The Sleeping Abductees*.

THE ZETA AGENDA

Through genetic engineering and inbreeding, emotions of the Zeta race were eliminated because unchecked emotions had previously destroyed their planet. It had taken eons for the Zetas to realize they had erred in eliminating emotions, but then they discovered that emotions are the bridge to the spiritual self. Without emotions they were unable to evolve in the spiritual realms.

* Anna Hayes is now writing under the name of Ashayana Deane

They had come to a physical and spiritual dead end and had lost their ability to get in touch with their spirituality. The Zetas lacked the genetic capability to be compassionate and benevo-

Draco/Zeta

lent, two keys necessary for spiritual evolvement. Needing to reclaim their emotions, the Zetas thought the best way was to observe and use genetic material from Earth humans. To acquire permission for abducting humans and experimentation, they made secret deals with leaders in Earth's governments to exchange their advanced technology for access to human genetics. The negative Zetas also wanted to gain control of the planet, and they made an alliance with the Dracos reptilians to try to control Earth. Conversely, the positive Zetas wanted to cooperate with the Guardian extraterrestrials to help Earth humans evolve. Multiple agendas developed that we will try to sort out.

The Zetas are a fragmented species according to Anna Hayes'* extraterrestrial source; their customs and policies differ among the various Zeta groups. Numerous subspecies have evolved from the hybrid program. Some are lizard-like in appearance with numerous colors and sizes, part of the Zeta/Draco hybrid program. The "little Grays" are referred to as the "lizzies, which are the Rutilia subspecies, reptilian in appearance". Some hybrids are spider-like, resembling our insect kingdoms. Others have been described as blues, browns, and silvers, who possess human-like or dwarf-like characteristics. There is also a white specie. The blue-skinned Zephilium are the administrator caste of the Zetas who govern the lower-ranking Zetas. Through genetic experimentation, many Zeta groups have resulted from mutations. As each successive generation developed, they lost the ability to naturally procreate their species and have become genetically neutered. Some sources claim there are 22 subspecies of Zetas.

The Zetas are intellectually and technically advanced, far more so than humans. They have no strong bonds of brotherhood with other life forms. With knowledge of the Three Dimensional Time Portal System, they have the ability to time travel. Some Zeta visitors are from the future and some from the past. These alien visitors are not inter-dimensional, but inter-time, understanding multi-dimensions to a high degree and frequently use the Time Portal System. Space-craft of the Zetas are three dimensional, with only the smallest craft operating in the Earth's atmosphere. With the ability to establish holding patterns between dimensions, they can become invisible at will. By using the Dimensional Lock System, they have the ability to appear in adjacent dimensions as well. The Zetas originate in the Third Dimension, but operate in a different time continuum.

Zetas are considered to be masters of the hologram. They can shape shift by modulating dimensional frequencies and have the ability to manifest in whatever form they desire, such as an animal or human. Most frequently the Zetas remain in the physical and use simple frequency modulation tactics to scramble brain waves of those viewing them. By interfering with the human bioelectrical system, Zetas can make themselves invisible, ghost-like, or disguised in other forms.

Our Secret Government has been working with the Zetas on research agendas, but Anna Hayes'* ET sources say the Secret Government is being manipulated to accomplish the Zeta agenda. They are a driven race, acting with one mindset originating from their home planet. They operate as individual units of a collective mind, none of whom thinks independently. One of their aims is to expand their reality and knowledge by consuming parallel versions of their own planet and to combine these reality fields with their own. The nuclear war on Apex had set in motion the death of their planet and its dimensional counterparts. By mastering time travel, the Zetas have gone back in time from the future to our time continuum to find a solution to this catastrophe in their past, because Earth humans contain the DNA of a civilization that could destroy itself with nuclear war. Because their time Portal System collapsed, they are unable to access their own portal directly and, therefore, entered

Earth's time continuum because our time portals are proximal to theirs. The Earth also has certain environmental elements the Zetas need – oxygen, water, zinc, plutonium, and iron-based minerals. By moving through the dimensional portals of Earth into a parallel system, they hope to access their own past planet and time continuum. By doing this, they hope to reconstruct the pathway to their own Three Dimensional zone to the inter-dimensional grid. Because of a lack of knowledge about the time portals in other dimensional systems, a large number of Zetas became trapped between dimensions and were stuck in our time/space continuum. They had leaped into our system from the Third Dimension by moving backward in time, but they have not been able to leap into another dimensional time continuum. Essentially, the Zetas are a dying race trapped in time far from their home.

WWII

Because of the carbon-based element system found on Earth, the Zeta's collective health was deteriorating. Once again they needed to remedy the situation through experimentation. About the time of WWII they designed a plan of genetic crossbreeding. They wanted to create a hybrid that would allow the continuation of their race, but the experiment was a failure with a variety of grotesque creatures being created. As a last resort, the human species was approached to create a Zeta/human hybrid that would allow the continuation of their race.

With the ability to shape shift, the Zetas were able to alter the human perceptual frequency enough to appear in whatever form they chose. Through these tactics, they were able to infiltrate the human government systems and tried to form agreements with the United States, Britain, and German governments. Not concerned with the outcome of WWII, the Zetas made a deal with the highest bidders – the Germans – in exchange for advanced technology. With their interest in genetic experiments, the Zetas were intrigued by the Nazi genetic experiments to create a superior race. The Zetas would allow the Nazis to help create a genetically superior human prototype that would lead to a Zeta/human hybrid. However, the Zetas became disenchanted with the Nazis because of their anti-Semitic policy. The Zetas regarded the Jewish race to be genetically superior. They wanted a human prototype that carried the code of the Jewish race. To protect the gene pool of the Jewish race, the Zetas began to negotiate with the Allies, using the guise of human forms. They wanted the Jewish genes protected at all costs. This interaction set in motion the seeds of the global Secret Government. The Zetas offered certain strategies and technology to the Allied military top command, seducing this human covert elite into believing the Zetas were far superior to humans intellectually. They had come disguised as Guardians under the illusion of being saviors of the planet. Military, medical, and communication knowledge was given to the Allies in order to manipulate them for the Zetas' own purpose. Following the war, the Zetas began to demand favors in exchange for their help during the war. Members of the covert Secret Government had no idea that they were being manipulated.

Post-WWII

Most world government leaders have little knowledge about the Secret Government and how it works. Some know of its existence but have been given misinformation. Those with direct knowledge of the Secret Government have agreed to perpetuate debunking the activities of UFOs upon insistence by the Zetas. Those within the traditional government who are somewhat in the know believe the Zetas and Secret Government are working with them to create healing modalities and economic strategies. However, the hidden agenda of the Zetas is to take control of the official government when the appropriate mass control devices have been set in motion. Official governments are fast becoming puppets of the Secret Government and are ignorant of reality. Part of the Zeta agenda is brainwashing through subconscious beliefs that most people have been taught not to question. The Zetas and Dracos depend on ignorance and disbelief to protect their covert agendas.

As discussed earlier, Keylonta is the language of symbol codes (subconscious) that create the foundations for all forms and structures within the dimensional system. It is a language of light, sound, pulsation, and vibration of energies. It also represents the living codes of matter and biologies that are built around them. Keylonta is the language of communication through the portals of time and dimensions.

Communications from extraterrestrials come to us from another dimension through a time portal system of an Earth that exists parallel to our own. The extraterrestrial cultures live within that parallel system, but in its own time portal frame. Keylonta is the means by which the Zetas are able to create perceptual interference and to direct and manipulate human perception, according to the Anna Hayes'* sources. They are able to manufacture perceptual interferences and sensual interpretations. They can create holograms and hallucinations. The contrived reality pictures can be coded by the Zetas to emerge as past memories or present Three Dimensional objective experiences. Events can be manufactured without permission or conscious knowledge. These reality pictures are called holographic inserts and have been used throughout history to alter the nature of our historical development.

The holographic inserts operate in several ways. By impulsing the base DNA code, the entire physiology of a person is altered on the chemical and hormonal level. As the neurological structures process the altered codes, individuals are perceived outside themselves as Three Dimensional matter – the images and events of reality pictures that were programmed into their DNA. This process can be used to assist someone with spiritual enlightenment, physical health, beauty, or mental or emotional expansion. Holographic inserts can also be used to perpetrate horrific traumas upon the biological system. Zetas are using holographic inserts against some individuals to mislead them, and these will later be increased to reach the masses. The Secret Government has allowed such experimentation and, ironically, the Zetas are secretly using the holographic inserts against the Secret Government in order to control them. The Zetas and Secret Government are planning a joint venture to manipulate and direct a large mass of people. Their plan is to create dramas using the traditional and New Age belief systems, such as true miracles and visitations.

Humans who are chosen to breed a work force to supply Zetas' needs often have the best genes to create and strengthen various hybrid beings to recolonize Earth. The Zetas are also attempting to find a consensual means of frequency through which they can broadcast the holograms designated to engage the emotional facilities of humans. Through the holographic inserts, they will be able to scan the intended events and create distorted versions of them. The Guardian extraterrestrials warn us that these renegade Zetas need to be stopped from fulfilling their plan.

The purpose of the Zeta agenda is to sever the energetic connection of a person from the personal Soul Matrix and to reconnect it to a group mind network existing within another dimensional structure, similar to the collective mindset that the Zetas operate under. Humans could easily be controlled because the false matrix would continually sustain certain holographic inserts. As an individual is separated from the original Soul Matrix, all the memory contained in that DNA and cellular patterns is wiped away. As the human is fully connected to the false matrix, the soul can be disengaged and the new matrix will program the DNA to download memories programed into the false matrix. If this is allowed to succeed, a person will be robbed of individual and collective memory, creating a biological puppet controlled by the Zetas.

This complex mental takeover has been accomplished at different times within all dimensions. Planets and galaxies also possess a Soul Matrix as do all humans. All these matrixes are created through and connected to the Time Matrix, which is the energy distribution system that carries life force energy from Source into the dimensional system.

Humans are connected by cords of energy to the Soul Matrix, which extends out to a sac of energy at its ends called the "nadis," which surrounds our physical body. Nadis contain all the life form energy that we will use in one incarnation. As energy leaves the nadis, it passes through two inner layers of the nadi structure called the emotional and mental bodies. We each have a double in the parallel universe called a "dolus." If the Soul Matrix is severed, the errant energy of the emotional and mental bodies will be rechanneled into our double, the dolus. The energy then becomes trapped within that dimension of the dolus and is not allowed to reintegrate with its Soul Matrix. If the cord of energy that once connected the individual to the Soul Matrix is disrupted, the Soul Matrix will become depleted if the energy leakage is not stopped. The energy would be drawn into the other dimensional Earth versions or parallel universe to the individuals who are our human counterparts.

An individual and its double exist as one identity package within the Soul Matrix. There are a total of twelve such identity packages within one Soul Matrix. Each of these identities are extended into different time/space continua within the dimensional system, with each identity experiencing its own line of linear development in its space/time continuum. These twelve identities comprise the soul family; each aspect or individual implies the existence of the other eleven aspects. Each aspect of the soul family has its own smaller cycle that constitutes its passage through time or its life span. A Soul Matrix cannot leave its dimensional octave until all of the residual energies are returned to it. If the Zeta matrix transplant plan goes into effect upon large numbers of humans, humans will become stuck within their positions in the Time Matrix, preventing their evolution until the trapped energies are released. A great deal is at stake with this human/Zeta drama.

One scenario to demonstrate how this mental takeover could happen is an appearance of a hologram of Christ. This "Christ" asks his followers to come with him. Masses of people would follow him through a dimensional portal, thereby evacuating much of the planet, freeing up real estate for the Zetas and Dracos. The Zeta plan is to use holographic inserts to orchestrate this mental takeover in order to gain dominion over Earth's territory, a plan that has been in effect for many years.

All religious texts have been subject to manipulation and distortion in order to control humans. Religion has taught people to disown their personal power and have programmed them to follow the religious doctrines of outside authority. This has disconnected people from their Host Soul Matrix. Religious teachings distort the teachings by making people believe

God can be found only outside themselves. The Sirians and Pleiadians have tried to protect us from the manipulation of other species, such as the negative Zetas. It is time for humans to take responsibility for their own protection in order to fulfill their genetic heritage.

Some souls who have incarnated on Earth, called "Starseeds," possess genes that have accelerated their path of evolution. The Zetas view Starseeds as potential breeders for their Earth hybrid-race. The negative Zetas fear the the Starseeds because they have the ability to see through the holographic insert. Starseeds are being protected by the Guardian extraterrestrials (Sirians, Pleiadians, etc.), while the negative Zetas are trying to keep the Starseeds from awakening. Both the negative Zetas and Secret Government are debunking holistic therapies, promoting drugs, repressing women and minorities, and debunking UFO sightings. The news media is also under their control.

The negative Zetas and Secret Government want to divide and conquer humanity, mainly through discrimination and prejudice. The Guardian ETs warn us that it is important to join with others in tolerance in order to prevent holographic inserts. These holographic inserts can be detected only through the use of high sensory perception by using senses beyond the five identified senses. Development of high sensory perception is part of our evolutionary path, and if we evolve to this level, we will be out of the negative Zeta mental manipulation.

According to Anna Hayes'* ET sources in *Voyagers*, the Guardian extraterrestrials "believed the threat of holographic inserts and forced, matrix transplant is a reality for the time period spanning the next 75 to 100 years." They say if the Zeta agenda does succeed, humans will be unaware as the majority of people will be transplanted into the Zeta collective Mind Matrix and programmed to see only the status quo. She was told the negative Zetas covertly control the directors who operate the world banking system and thirteen top corporations, the One World Order, which is already in place and will function as a global government once events play out. In the new society controlled by the Zetas, there will be a large increase in emotional and psychological problems that will be controlled by drugs. The high achieving hybrids will infiltrate society and begin to exterminate genetically undesirable humans.

On the other hand, if this negative Zeta plan were abandoned or stopped, there would be an increase in UFO sightings, alien encounters, and communication with extraterrestrials. Tolerance would increase, and advancement in holistic healing would manifest. Freedom will be the purpose of society. Through intuition, we have the ability to sense manipulation by the negative Zetas and Secret Government. By directing the energy of the Soul Matrix into our conscious mind, we can avoid the influence of the negative Zetas.

The New Agenda

The Guardian extraterrestrials are very concerned about the negative Zeta agenda so have come up with a plan that might solve the problem. They want to relocate the Zetas to an adjacent dimension where they will be welcomed. Humans are asked to help broadcast telepathically to the Zeta community regarding this new option. This will help them align to one of the inter-dimensional grids assigned to them. The adjacent Earth dimension, to which the Zetas are being directed, lies in the Fourth Dimension. It is inhabited by humans, extraterrestrials, and inter-dimensional inhabitants. The Zetas will need help to integrate themselves into the system.

ABDUCTIONS

The Zetas are constantly experimenting to gain knowledge that might save their species from extinction. Because of their sense of urgency and the wide availability of human DNA, humans have made the perfect "guinea pig" for research. In the 1950s, the Zetas made a secret agreement with the U.S. government to allow abductions of U.S. citizens and experimentation upon them in exchange for advanced technology. Under this agreement, the abductees were returned to their place of abduction without memory of the abduction experience. This sounded like a sweet deal for the government, but the Zetas have abused the agreement and expanded their protocol.

There are two primary reasons for alien abduction. One is for abductee training, as many humans have agreed to participate in the abduction process. Most agreements are made during the human dream state through the subconscious. Many agreements are arranged in a soul agreement prior to birth, where the incoming soul chooses to experience the abduction event. The agreement provides that the abductee will not consciously remember the experience, but subconsciously be motivated to discover the higher purpose of why the soul incarnated. Once the emotion of fear subsides, the human abductee often embarks on a path of accelerated spiritual growth and awakens to soul identity.

The second reason for abduction is for experimental research and genetic seeding by self-serving groups such as the negative Zetas. These abductions are not conducted with consent, but harm is not deliberately initiated. Those humans who have worked to build a conscious connection to the Higher Self and have a healthy flow of soul energy will usually not encounter these abductions. Abductions that occur without soul agreement are a violation of human rights and can be traumatic, causing emotional and mental difficulties.

Interaction with the Zetas occurs on three levels – physical, dream state, and quasi-physical. The Zetas believe they have soul permission to conduct the abductions and research. They view our human collective soul as being part of them and, therefore, by the concept of unification, permission has been granted. In most cases individual agreement has not taken place, but there are cases of individual agreements. The Zetas say it is an agreement among our collective souls. They usually interact with us during states of altered consciousness when individuals are open to agreement. Often a signal is emitted and those who wish to participate will signal back. The Zetas view humans as fragmented, because we do not have awareness of what occurs on other levels of our own consciousness. Trauma of abduction occurs only when information has broken through the barrier between unconsciousness and consciousness.

When individuals incarnate on this planet, they have a signature vibration that is coded within the Zeta computer, and technology can determine an individual's development stage for particular experiments. Through a homing device, Zetas are able to locate that person, which automatically transports the Zetas to the space/time where the individual is located. They also broadcast beams of encoded electromagnetic energy that request volunteers for contact.

Zetas exist in interdimensional space, but with the flip of a switch they can enter our reality and time continuum. The space-craft to which the abductees are taken awaits in the Third Dimension. There is a vast network of space-craft where the operation is coordinated, and this is where the Zetas live, as they do not return to their home planet. Crews spend most of their lives on board the ship.

Zeta brains actually form part of their vast computer system through a process based on harmonics. When they generate a harmonic frequency, they are linked into the whole mass. The main brain system of the computer is actually organic, which is an extension of the inorganic components with which they can interface. Because the Zetas are linked to the mass whole through the computer, they consider themselve as "one people."

During the abduction process, mostly at night, the space-craft translocates to the homing signal. Once the individual is in the appropriate state of consciousness, the Zeta abductors automatically appear. The mass mind computer does all the calculations for the precise moment of abduction. Once the individual is located, a mass mind energetic field places the human in an altered state of consciousness. The person is immobilized through paralysis to prevent abductees from injuring themselves, and this paralysis is initiated by an electrical charge to the brain generated by the mass mind computer.

A translocating device alters the abductee's molecular vibration from matter to energy and a shift in the space continuum transports them onboard the craft. The process is then reversed and the individual is resolidified into matter. Another method to bring abductees onboard is by reversing the polarity of the body in relation to the Earth's gravitational field. The abductee is then guided physically onto the space-craft, somewhat like pulling a balloon on a string. Humans can pass through walls during the abduction experience by Zetas altering the molecular vibration of abductees, making them less dense than the solid object. Of course, the abductee is unconscious.

Onboard the space-craft, the abductee is brought to an examination room. Zetas can communicate with abductees only when they are in an altered state of consciousness, not when the abductee is in a conscious stage. After the abductee is placed on a table, an inorganic probe is frequently inserted into the brain, often through the sinus cavity, nose, ear, or eye. Most probes will collect and catalogue neurochemical data. Over a period of time, probes are removed and the data recorded. Probes are composed of condensed plasma light energy that can be dissolved at will if there is a chance of detection.

Often the probes are placed in the anal region, analyzing fecal matter to give data about the human digestive system. In males, a probe is used to extract secretions from the prostate gland. Some people report afterward about painful irritation in the areas of the probes. It is estimated that 20 percent of the Zeta crews are clumsy in their protocol and may cause discomfort to the abductee. Most often, the Zetas induce a temporary memory veil in the abductee because a conscious memory will not be of service at that time. It is easier to block memories when the person is not conscious and in a dreamlike state. After the examination, the abductees are returned to the pick-up point, which is usually their bed. Zetas rationalize that they are helping the abductee evolve by generating in them their deepest fear.

The Science Underlying Abductions

UFO abductions are accomplished through a process called dimensional transmigration. This is done by form transmutation, when an object in one dimension is changed by moving it into another dimension for alterations. It is then returned to the original dimension, where the dimension alterations appear as a change in form. Humans are temporarily neutralized and the human mental awareness becomes momentarily disengaged from the body. It is put on hold when the body is taken into a dimensional frequency or through a time portal. It is as though the astral body is pulled out of the physical body. In reality, the energy fields of the mind and that of the body are put into different vibrations. The mind is accelerated and the body is deaccelerated. When the proper ratio is reached between the body and mind, the body is quickly accelerated and leaps through a time or dimensional portal. Thus, a window of time is created, allowing the body to move through a portal without its form being destroyed.

During an abduction, the physical body is taken but is shape shifted and literally turned into light formation in another time and dimension frequency. First, it is taken back in time to its prematter form and then thrust into its own future. It is then moved from the future perspective backward toward the present to reemerge with the mental energy that was put on hold. Alterations to the body can occur when the body is in nonsolid form, existing in light and sound. When the body and mind integrate, these alterations take on a symbolic code that gives the illusion of form. To the body and mind, these events will appear as if they had taken place physically because the physical sensory imprint is valid within the present form of consciousness. The remembered events are real, but actually took place in the future where matter had not caught up with them. Memory is stored within the body cells and DNA, which is where the events of the future are recorded, consciously remembered in the mental body.

Scars may appear on the body as the body reintegrates the cellular and mental codes. Implant devices may be found in the reintegrated body and fetuses may be missing. The illusion of what you see as "cause and effect" is a primary base code of the Three Dimensional system.

Because of Genetic Time Codes, all the frequency bands, which actually exist at once, are perceived to be strung out in a linear fashion. Our consciousness is fragmented into various frequency bands, but reassemble into the frequency of the present moment. The Genetic Time Code sorts and organizes the electrical code experiential data into a pattern that the conscious mind can synthesize into a sequential reality.

Abduction experiences are quite real. When the transmutation process occurs, the mind will have no recall of the event.

Missing time is experienced as the body and mind reintegrate these vibrations. It takes time for the conscious mind to catch up with the altered body imprint. When the mental faculty integrates the new imprint, flashes of memory will come into conscious awareness triggered through association, like smell or sight or sound. Such memories can also be triggered through hypnosis.

Some abduction cases involve the tactics of perceptual interferences, when hologram inserts are used to cloak the abductees' perception. Abductees may remember seeing eagles, bears, or humans instead of aliens and spaceships.

Abductees who remember parts of the abduction encounter frequently have emotional problems trying to sort out the experience. Fortunately, there are a number of psychotherapists around the country who use hypnosis to help the abductees deal with their abduction experience, including psychologist Leo Sprinkle, Ph.D., professor emeritus at the University of Wyoming, who has a private practice in Laramie.

HYBRIDS

Many abductees report the Zetas extracting sperm and ova from them while they are on the examination table. The specimens are used by the Zeta to create human/Zeta hybrids that the Zetas hope will preserve their race and be adaptable to Earth's conditions. Experiments are conducted on both the physical and etheric dimensions. The etheric dimension exists outside the physical level. The physical matter lies adjacent to the etheric state and forms itself according to the light plasma language coding of the etheric. The Zetas are capable of creating a template from which a physical life eventually forms itself. They insert the plasmic light energy with encoded language into the template that instructs physical matter to create a body structure. The encoded language first affects the nonphysical and then the physical, which is arranged according to the etheric template. Zetas have technology that can alter the light plasmic language to find the most perfect body structure. What is done etherically will affect the physical. When the etheric template is given enough plasma energy, it can then enter the physical realm.

After a fetus has developed, the Zetas can monitor and direct its biochemical development so it will have the correct proportion of biochemicals. Fetuses are either implanted or naturally conceived. Between one and four months the fetus must be removed from the human embryonic environment. The most frequent procedure is to vibrationally alter the molecular structure of the host mother so that it becomes less dense, so the Zeta surgeon can insert his or her hand into the womb and remove the child without blood loss, much like a psychic surgeon. A less common method is to remove the fetus through the vaginal tract.

Sperm samples provided by human males are often genetically altered in the laboratory. The Zetas have vast sperm banks collected from humans. Sometimes they will abduct males so they can give emotional support for the females.

Human females who made a subconscious soul agreement to provide the eggs to foster a hybrid do not remember the abduction except under hypnosis. Their genetic codes often do not allow many of them to engage in interdimensional transmigration, meaning they do not have the gene of transmigration. Their extracted eggs are used to create the white hybrids. Another group of women experience missing fetuses called vanishing pregnancies. These women serve as surrogate mothers. As the hybrid fetus develops, the vibrational pattern in the genetic base code begins to change. Because the fetus has become incompatible with the host body, it must be removed to another dimension where its development can continue. Only white hybrid fetuses can be taken, and these hybrids cannot be brought into the Third Dimension. Sometimes visitations are conducted between dimensions where the human mother can meet with her hybrid child in a neutral surrounding.

After the child is gestated, lasting 10 to 12 months, the abductee is brought to the incubation chamber to interact with her child and give it love. Hybrid children have surrogate caretakers who interact with them daily by giving them comfort and acts of love. Most hybrid children do not have a chance to bond, and most do not live long enough to reach maturity. Those who survive beyond childhood are very weak. Overall, the hybrid program of the Zetas is not close to creating hybrids that can populate the planet. Lyssa Royal was told the hybrid experiments should be completed by the year 2000. All the hybrid beings the Zetas work with contain Zeta souls that have chosen this experience. The hybrid program is symbolic of a marriage between our two species and represents the future of both humans and Zetas. The Pleiadians often work with Zetas and their hybrid program.

With all their research on hybrids, the Zetas still have not achieved a stable hybrid strain, mainly due to a dysfunctional immune system, as many of the hybrids become sick and die. They have also been having a difficult time developing a strong hybrid skin, pituitary hormones, and growth regulation. Often hybrids develop symptoms of radiation sickness. By 1990, only one hybrid had been able to conceive and give birth to a live child. Emotions have been another problem area. A few have experienced laughter and compassion, but few hybrids have emotional experiences. No hybrids have felt loneliness.

While the positive Zetas are trying to help humans and their own species, the negative Zetas are trying to create a hybrid whose ultimate purpose is to compete with humans for Earth and its resources. Some of the hybrids already exist but are not quite strong enough to be a threat. Zetas have successfully interbred with human females, resulting in babies, children, and Zeta/human hybrids. However, these hybrids cannot enter the Third Dimension as their biology cannot adjust to our matter density, so they exist in an adjacent dimension band of Earth. These hybrids are being cultivated to interact on Earth, perhaps 400 to 500 years in the future.

Zionite (Hybrid)

There are a variety of Zeta/human hybrid strains. One strain is large, with wispy hair, and white in color. They possess the capacity to understand human emotion better than most strains. Another strain called blues encompasses dwarf-like creatures, while another strain is brown, created to be workers. They represent a conglomerate of genetic material derived from other dimension species plus human DNA. These beings do not have high intelligence, but are programmed to serve the needs of those on a higher evolutionary scale. Often they serve as guards.

The highest strain, called the Zionites, is a genetic conglomerate of Zeta, human, and Aetheian (mantis-like) beings. This strain constitutes a master race whose power reaches far beyond the present moment of Earth history. The Zetas drew genetic material from the distant past of our species and sent mature hybrids back in time to directly interact with humans. Zionites are great time travelers.

Aetheian

Zionites reside on planets in our universe and many work with the Aethians in their home system. They exist as part of our future in Third Dimension reality and have had a tremendous effect upon our species and are part of human heritage.

The positive Zetas are also giving something in return for the abduction experience. They are giving us more codes that activate the light plasma language level, as thousands of people have experienced an abduction. When a critical mass is reached, the human species will leap forward in evolution. We are in the last 25 years of achieving a global vision of becoming citizens in a global civilization. Some of the genetic codes the Zeta are triggering allow us to see ourselves as part of the galactic civilization, a working piece of the whole. One of the genetic codes valued by the positive Zetas promotes unity consciousness rather than diversity. They are putting this into the etheric template. From us, they took the genetic code of individuality and our love of emotion. Eventually, we will be encoded to symbolically speak the language of our galactic visitors.

EMOTIONS

Many Zetas fall behind humans regarding spirituality because they don't possess the physiology that allows emotions. In Anna Hayes'* book *The Voyagers,* she relates from her ET source that "Emotion is the product of the seventh sense, which is called absorption." It is the ability on the cellular level to filter energy and then create energy imprints that translate the imprints into perceptions that are fed to other senses. Originally, the Zetas possessed emotions similar to humans, allowing them to connect with the universe grid structure that supports the Third Dimension. Through genetic engineering and mutations, they lost their ability to express these emotions. They engineered out the physical and neurochemical reaction that produces emotion. They then realized they had made a mistake and essentially became trapped within their own bodies and their souls became disconnected from the Soul Matrix. The Zetas are able to draw energy from their souls, but not from the Time Matrix, Energy Matrix, or Source. As a result, the Zetas are a dying race unless they can genetically engineer their species back into the Time and Energy Matrix system. Because of their ignorance and desperation, the negative Zetas use methods that are contrary to spiritual evolution. The enlightened Zetas are beginning to understand the Law of Cause and Effect and are working with the Guardian extraterrestrials under the policy of nonviolation.

Zetas believe that fear is a vital part of human transformation. By releasing fear, humans can transmute the emotion of fear, which can be done by visualizing a solid blue light around our body that makes our field dense. The Zetas believe that through abductions they generate in humans a method to face their deepest fear, which will help the abductees transcend to another level of evolution.

Conversely, the negative Zetas feed off the emotion of fear. Once a person becomes fearful, they have you. Fear puts the heart and mind out of synch. This causes a person not to think clearly or be in focus, which enables the Zetas to control him or her. Those who have explored their psychic potential have a much better chance of avoiding control by the Zetas, as compared to those who are grounded in Third Dimensional reality. When the paranormal becomes normal for us, the negative Zetas will have to search elsewhere for their nourishment from fear. Fearful people attract fearful contacts, and those without fear have a better chance of being left alone.

The Zetas are trying to understand the link between emotions and sexuality. Presently, there is no physical or psychic difference between male and female Zetas. Because they thought that sex had contributed to their demise, their sexual energy had been channeled into other areas, especially mentality. Their minds are totally fused. Now they are realizing that sexual behavior can help them express emotion. Because emotion is so involved in sexual activity, Zetas often observe humans having sex. This explains the Reptilian Dracos wanting to be present during the act of rape to feed off the emotions of fear during a rape and the sexual emotions of the perpetrator.

The hybrid Zeta has simulated sex organs and is able to mechanically have sex with humans, but cannot create any physical sensation through their simulation. At this stage, they are not fertile. For those Earth females who have been impregnated by a Zeta, the child could not live on Earth. The Zetas have been trying to understand the concept of love but have not allowed themselves to love. They are hoping to give themselves neurochemical substances that will trigger emotional responses, such as love.

The connection to the whole is what motivates the Zetas, who are like a mass mind, as compared to Earth humans who are individualistic and are not necessarily connected to the whole. Perhaps humans can learn something from the Zetas about the importance of the whole while the Zetas learn from us about individuality and the creation of emotion and love.

Chapter Six

THE ARCTURIANS
Our Future Selves

Of all the galactic civilizations that have been assisting Earth on its evolutionary path, the Arcturians are probably the most evolved. They have played a major role for eons in guiding Earth's civilizations. Arcturians are Fifth Dimensional beings from Arcturus, which is the first brightest star located in line with the Big Dipper handle located in the Boote's constellation. The Arcturians' focus is mainly the inner world, aided by ascended masters and Jesus, whose galactic name is Sananda. Their mission is of love and light. Arcturians never tell Earth humans what to do or what to think, they only suggest. They never criticize other civilizations or individuals. Their message, like all the galactic civilizations, is that Earth will be entering the Fifth Dimension, and they are here to assist us in the transition. They can teach in the Third, Fourth, and Fifth Dimension, and they will give us universal law that will apply in the new millennium as well as in higher dimensions.

Their purpose is to help Earth make the transition into this new era. They emphasize that the key to entering the Fifth Dimension is love that will enable us to reach this higher frequency. In order to do this, negativity must stop. The emotions of fear and guilt that so many humans possess must be exchanged for qualities of love and light, which will enable us to have peace, harmony, and ecstacy. Their mission is to fulfill the plans of the ascended masters led by Jesus the Christ. The Arcturians have a plan to save Earth from its own possible destruction.

They have used Dr. Norma Milanovich to help convey the message about the upcoming transition. Norma was an assistant professor at the University of New Mexico when her life changed forever. During the 1980s, Norma began to develop psychic abilities, which she couldn't explain with traditional science. One day she heard a voice say *pick up a pen*, which she did and proceeded to watch her hand move, forced by an energy not of her own. This happened long before she had heard of the terms of automatic writing and channeling. Eight months later, Norma began to receive messages on her computer from an extraterrestrial group called the Arcturians. Much of this chapter is based on her experience with the Arcturians that she wrote about in *We, The Arcturians: A True Experience.*

All messages transmitted from the Arcturians were recorded in a form nearly identical to those received during transmission. When a question was asked of the Arcturian Being, Norma would enter a semitrance state. This allowed the Arcturian source to communicate though her by way of a microcomputer. Once the Beings began sending the information, Norma received it telepathically as fast as they spoke. She then began typing the message nonstop on the computer.

THE ARCTURIAN BEINGS

Wanting to have Norma understand who they were, the Arcturians described themselves, their planet, and their philosophy, hoping this would assist in the birthing of a new era on Earth.

Physical Characteristics

Norma was told that Arcturians were short in stature, about three to four feet tall. They were slender and all looked alike with various skin colors, mostly gray or green with a beige tinge and white tone. Their two arms have hands with only three fingers. They do have legs they use on journeys, but they don't need them because they can float or levitate. Because they are on the Fourth and Fifth Dimension, there is no such thing as gravity. They direct their motion and move in that direction by tapping into the consciousness of liquid light. They mainly use their legs for needed exercise. They have no ears, hair, eyebrows, or eyelashes. They call the mouth "the oracle of truth," even though they don't communicate with it. It is an opening used for ingestion. Norma was instructed on how to draw them, as she found that the Arcturians had command over her thoughts and equipment.

For nourishment, they consume energy, preferring mainly effervescent liquid that is high on the vitality scale for spirit. Foods that vibrate on the lower level cannot be consumed by the Arcturians. Lower vibrational foods dull the senses to higher consciousness. Their space-craft can produce almost any combination of vibrational food, with flavor and odor not important to them. After running low on a needed vibration, they can concoct a high grade of energy through vibrational food that accommodates their spirit, but only when necessary.

Arcturian

Their almond-shaped eyes are dark brown, almost black. Dark eye coloring helps shield their eyes from certain harmful rays of the sun, which diminish their telepathic abilities, their primary mode of seeing, as eyes serve as a secondary input source. The Arcturians told Norma that the eyes serve as sensors for the "spiritual nature of our beingness." They are the windows of their souls. Eyes can also filter out lower vibrations that affect the quality of life. Behind the coloring of the eye is a sensor that can perceive distance in very minute detail. The data collected by the sensor is transmuted back to the central nervous system to give the quality and character of what is seen. The eyes have the ability to have inner and outer vision at the same time.

The Arcturian's heart holds the answers to universal keys, Norma was told, as it can balance both positive and negative energy. Their mission on Earth is to teach humans to hear the energy source and the words that come from the heart.

Metabolism for the Arcturians is much higher than humans' allowing them to find enlightenment in everyday affairs. In the future, they told Norma, the metabolism of humans will increase as the level of consciousness increases.

Arcturians no longer have ears, but five small holes on each side of the head in a shape of a pentagon for the purpose of ingesting certain frequencies. They have an inner eye and also one for hearing. Their sense of hearing transcends even their telepathic ability. Sounds can be heard from the Fifth Dimension and higher.

They sense with the back of their heads. The tilt of their head enables them to gain access to different forms of information to help make the decisions they wish to manifest.

Reproduction in an enlightened being is assigned to a chosen few. Arcturians don't mate physically, as they are truly connected with the spirit of the holistic concept of God. Some Arcturians have vibrational frequency patterns that are conducive for reproduction. Within one of two members, the cellular structure begins to sing the music and vibration of a new life, meaning the Arcturian is pregnant. They reproduce with pride.

An average life span for an Arcturian is between 350 to 400 Earth years. By not acknowledging time and age, Arcturians can transcend them. When coming into life, they arrange a contract for existing. They anticipate the length of their contract by the number of accomplishments they want to fulfill. This is measured by the vibrational level they are able to attain. Many centuries ago, sickness was eliminated on Arcturus. They control their population by planning for the highest in their evolutionary patterns. If something goes wrong, like a space-craft crash on another planet, the dead Arcturian is excused from its assignment, but reports back to the Elder in another form of consciousness.

Clothing

Away from Arcturus, they wear bluish green clothing without pockets or openings, designed for protection outside their own solar system. The clothing fits closely to the body and serves as a filter of the consciousness and thoughts of those who interact with the Arcturians. To move out of the garment, they pass from the Fifth Dimension to the Fourth Dimension, and on to the

Third Dimension. Clothing is mainly used for protection in the Third Dimension. A special chamber on the craft provides the technology for dressing and undressing.

On their own planet, they wear free-flowing garments to not hinder locomotion. Sometimes they wear uniforms for protocol, and some garments are worn mainly for uniformity of respect. Clothing is not worn on Arcturus for warmth but only for protocol and respect.

Families

Traditional Arcturian families are quite different. They have what they call the unit, which plays a dominant role in controlling the youth's vibrational frequency. The unit is composed of wise souls who understand the principles of energy. The wise souls selected at birth to be part of the unit are of a frequency similar to the color violet. The color of the aura reading tells the potential for that soul's accomplishment. Souls are selected on their color frequency, and only the wisest are allowed to associate with the young. These wise souls can promote increased frequency of the young at a faster rate. Those who govern the units are the highest and wisest in the land.

Recreation

Arcturians believe that all of life is recreation. They know only oneness and not the strife that comes from duality like that on Earth. Arcturians have patterns of mind control and games they enjoy playing. They delight in all mental feats, as the mind takes control over the physical as civilization progress. They enjoy creating and have no need to want anything.

Time

Arcturians are outside time and space. The only measurement they honor is measured by frequency counters of the soul. They don't have minutes and hours, but increments of evolution. They measure only the degree to which they are evolving.

Arcturians are in charge of their realities and can manifest anything they want to occur. At the same moment of time and space, two Arcturians can manifest two opposite kinds of seasons.

Atmosphere

Arcturus has two suns that create a reddish-violet light that surrounds the planet. Their atmosphere is made up of various gases that have interesting functions. One gas in the atmosphere prevents damaging forms of microorganisms from multiplying. Another gas is for breathing, and a third gas is watery in nature and supports the ingestion process. The most prevalent gas in the atmosphere cleanses negativity. All the combined gases form a haze around the planet.

History of Arcturus

The Arcturians claim the soul of humans was cast at the same time as the souls in Arcturus and other galaxies. At one time eons ago, the Arcturians had established a colony on Earth. During this period there was a warlike consciousness that destroyed much of the civilization. Many of the colonizers were rescued and taken back to Arcturus. Those left behind were given the privilege of reincarnating in other forms. The Elders made a decision that it was no longer feasible to return to Earth and worked with a nearby galaxy to set up a tribunal council to arrange for the first inhabitants of other species to journey to Earth.

The Arcturians told Norma the energy around Earth makes it difficult for many souls to stay pure and remain at a high level of vibration. In the past, life forms took part in mating with lower life forms, and space visitors interbred with many of the human species and left qualities of godhood with them.

Concept of God

The Arcturians believe there is only one God and one eternal life. God is the force present throughout all the universe and is the light and intelligence of this world. God is centered in the Great Central Sun.

They say God is manifested in many different forms of mental energy. Until each soul finds God, a level of discomfort arises. Love, light, and electronic forces are all God. Within His Beingness, they are all one. However, within human consciousness, they are three separate things. Love is the power that holds all together on the emotional level.

Great Central Sun

The Great Central Sun is God, the highest vibrating substance in all the universe, the expression of all that there is. If one looks within, one can find the Great Central Sun there and will have access to all universal knowledge gained from the journey.

When a consciousness comes into the world, it will experience separateness and will have a difficult time relating to the Oneness. Duality is part of consciousness, and souls begin to feel that all degrees of differences are not opposite or in competition with one another. It experiences darkness as one aspect of light on its path to Oneness.

As the curriculum of the God force is learned, a soul is ready to transcend to another dimension and learn higher lessons. As one accepts the basic understanding of Oneness, the soul finds no conflict anywhere, for in the acceptance of Oneness, a soul sees perfection in all that is.

By completing so-called evil acts, one learns how to raise his or her soul's vibration. When a soul sees the goodness of God in all, the soul is ready for a higher dimension, for the Earth plane is no longer a challenge. God is both light and dark!

Dimensions

The Arcturians described to Norma the differences between the various dimensions. In the Fourth Dimension, some souls got confused and actually thought they were on Earth, a planet of duality. There are creatures of many universes that exist in this dimension that find a harmonious state on their path to a higher level of existence. The two major qualities of the Fourth Dimension are love and forgiveness as taught by Jesus. Once these are mastered, the vibration frequency is altered to prepare for the Fifth Dimension.

The Fifth Dimension has a vibrational quality like "heaven." Anything a soul in this dimension desires will manifest using the force of love. All the beings are androgynous and live in totality of this Oneness. To enter the Fifth Dimension requires opening the heart center. On both the Fourth and Fifth Dimension, one can learn the path of soul evolution. The Arcturians say that Earth will be grounded in the Fifth Dimension within the next 50 years, as told to Norma in the 1980s.

In the Sixth Dimension, the rays or emissions from a Sixth Dimension being are so intense that they can disintegrate the highest frequency known to Earth. Color and sound frequency are used in this dimension. On the Seventh Dimension reside ascended masters who can access all the dimensions. The Arcturians say they are only one step ahead of Earth humans in leading the way on our evolutionary path.

Communication

The word "language" on Arcturus implies the energy that is formed within one's consciousness that directs the visualization process. Both the encoded message and intended feelings are communicated, with the word message carrying energy to the appropriate section of the consciousness. On Arcturus, they allow symbols from the universal mind to permeate consciousness. These universal symbols carry information from the source. The Arcturian communication system provides more accurate transmission of messages in both content and quality of emotion.

Communication between various species is often telepathic, while the language system on Earth, the Arcturians claim, sustains duality and not the desired holistic pattern. Visualization patterns are the key to our survival in the New Age, providing the foundation for accurate communication.

Arcturians do not have a brain structure that is common to humans. Their brain is composed of a complex system of fibers and neurons that control the frequency pattern of their essence. They have access to universal language codes, symbols, vibrations, and frequencies. By moving these symbols to different dimensional frequencies, they can form these into any pictures or visualization they wish. The reason the Arcturians are so telepathic is that they have the ability to access any of the symbols in another being's consciousness. They also have the skills to access information from the collective consciousness.

The Arcturians say "it was a lack of emotional retardation in their evolutionary progress for eons" that created many problems for them. Realizing they needed to change, they began to study the universal coding for peace and harmony. The codes vibrate on a higher frequency and rest within the capacity of each Being. After incorporating these codes into their lives, their evolution accelerated. Much of what the Arcturians learned came from the advanced star system civilization Alpha Centauri, which had merged with many of the ascended masters.

The Arcturians discovered that the frequency at which a being vibrates is directly correlated to the command one has over thought, action, words, and emotion. When one's frequency increases to the speed of light, it can access information in the universal consciousness. Beings with a high frequency are protected from receiving lower frequency energy because the light consumed within the essence of that soul is impenetrable. When a being is centered, the energy pattern is more logical, holistic, and not randomized. When centered, one can access the universal code, and when tapped into this frequency, one can understand another frequency pattern. Being in a state of Oneness, Arcturians are able to transmit and receive messages in perfect understanding.

Arcturians have the power to hear all that is being discussed telepathically. In order to accomplish this feat, they had to learn the techniques of clearing the mind. A clear mind is essential in their mode of telepathic communication. It is difficult for humans to clear their minds because of ego and their competitive nature. Competition produces aggressive behavior, rather than passive, which is needed in a clear mind. To become passive, a soul must learn to increase its vibratory rate.

Thought Transmission

Everything is energy, and all thoughts are electrical impulses surrounded by an electromagnetic field of energy that moves at extraordinary speeds. During this energy movement, images are transferred that are directly related to visualization, emotion, and the coding of the universal language. An individual has a choice whether to send electromagnetic impulses to another place. The choice is determined by the amount of will power and emotion that an individual can gather. An individual determines the degree of mastery over telepathic communication he or she obtains in one lifetime. The force that propels thought is based upon a tensor equation composed of two parts electromagnetic energy and one part emotion.

Norma was told that all humans think and behave in a linear fashion. However, there are no "points" in the universe of Oneness. Oneness supports only the concept of a holistic mind, meaning the mind is all "point" inclusive. Each "point" contains the Oneness of the universe.

The force that projects thought forms is electromagnetic energy of the universe, which is the yin and yang or Prana of the life support system of the universe. This force is the positive and negative, the light and dark side of our consciousness. It is both the subatomic and the void.

The degree of manifestation of the thought can be determined by the tensor equation. Each thought is part of the universal consciousness. The energy used to move a thought through one's consciousness is the intensity of conviction of needing to be one with that power. All thoughts that can ever be created "exist," as everything is already in the past, present, and future.

All manifestations of creative energies are the result of combining light energy and sound vibration. Many scientific discoveries are based upon the combination of light and sound wave. By combining light and sound energy, one can channel the power to create from the universal consciousness and access its codes.

The Arcturian evolvement is dependent upon their abilities to access the universal codes collectively. They describe their civilization as peaceful, not happy or sad. They have great contentment and peace of mind with a sense of accomplishment. They attribute this peaceful nature to their ability to master universal symbols and telepathic communication. The Arcturians are sharing this information with humans as they had received it from Alpha Centauri during their earlier stage of evolvement.

Education

The Arcturian's system of learning is based on higher states of consciousness and the vibratory frequency of the universe. Being of Fifth Dimension, they are total telepathic beings who can manifest objects with their minds. The light frequency of each Arcturian Being is the true measurement of that soul's evolvement and success, not facts and data.

The Arcturians are concerned with future development and have a hierarchal structure that directs knowledge and learning. Instead of teachers they have Elders. They do not have any grade level on their planet. Instead, they have what they called "language increments," meaning that the rate at which an entity vibrates is continually being monitored. As their frequency reaches a certain level, the learning is then increased to the next increment. There is no competition with others, only themselves.

There is no standard curriculum in their education system. All knowledge from the collective consciousness is offered, and each Being has equal access to information and knowledge. They accelerate at a rate they find enjoyable.

Frequency levels are the measurement of enlightenment, and enlightenment is the measure of true growth in intelligence. What is learned in one increment is used for the good of the planet, as they are the planet of One. They say duality is reserved for Earth. The Arcturians flow with the force and don't fight it! As a result, they have no stress. By going with the flow, they increase both their life span and knowledge.

They use their mind to create, not to destroy. Creation is learned when they channel love, and love is channeled to all their counterparts. They are the masters in the structure of learning. The Arcturian star fleet's main purpose is the curriculum of the mind. They are here to help Earthly humans learn the truth of life so we can evolve to the Fifth Dimension.

Enlightenment

One of the important principles the Arcturians use to help enlighten the mind is to remove all blockages that might hinder smooth energy flow through a life form. Fortunately, Arcturians do not learn about negative forms and frequencies of vibration in their society. They have several methods that reverse any blockages:
1. The first is to work on being open-minded.
2. The second is to understand and appreciate God, as all knowledge comes from the God force. On Arcturus, at the moment of conception, knowledge is implanted in their consciousness.

3. The third method is to channel unconditional love. The amount of love a soul channels can be measured by a frequency count. Knowing the Oneness is where they are tested. Increasing enlightenment increases power in the realm of creation and manifestation. Arcturians refer to enlightenment on Earth as increased intelligence. Power increases one's enlightenment. Elders help the process by concentrating on the picture of perfection, which helps the individual to raise its consciousness to that level.

Evaluation Process

Evaluation of an Arcturian's development is done by the Elders. They are only measured against themselves and can request an evaluation regarding the usage of their manifestation powers to qualify for a higher rank. A subject is placed in a room with Elders who can read the Akashic Record and document the areas of growth and achievement the individual has accomplished over eons of time. The Elders and Cosmic Beings determine from the Karmic Board level who is deemed suitable to progress to a higher state of being. During the exam, the individual is asked to resolve the unresolved situations from past lives that are recorded in the Akashic records. They are asked to provide the solution for unresolved situations.

On their spiritual path, Arcturians do not go back in vibration. They remain at the same level or progress forward. Arcturians are born into this higher frequency and contain a consciousness that transcends the Third Dimension. Only when the three bodies (physical, mental, and emotional) are in harmony will the individual begin the transformation into a light body and make a quantum leap into the new dimension.

Arcturus

The vibration emitting from Arcturus affects the entire universe and is designed to penetrate the consciousness of all beings. Arcturians broadcast from two frequencies, one from the energies of the planet itself and the second from their space-craft. The planet of Arcturus is a sacred dwelling of the highest form of life the universe is capable of producing in the Fifth Dimension. Arcturus is the teacher.

Arcturians have a natural abundance of positive energy, which they call the universal fuel supply. The planet was like Earth before it evolved into the Fifth Dimension, and they can't understand why we don't take better care of our planet.

Arcturus is a sister star in the heavens for Terra. They are the caretakers of the Fifth Dimensional inhabitants, while earthlings are caretakers of the Third Dimension. Earth is considered the outer world of reality, while Arcturus is the inner world of reality. On the soul level, Arcturus is the soul existence of light, while Earth is the light manifestation of the soul.

THE ARCTURIAN MISSION

The Arcturians told Norma that their purpose is to assist Earth as it enters a new era of spirituality. They are here to help educate and raise the vibration of those who choose to journey into the new dimension that we are about to enter. They will not interfere with the free will or decision-making of any Earth human. Their mission is to help us understand the nature of God, ourselves, and the universe. Every soul that incarnates has a purpose or soul destiny, and everybody has a separate path. When all these paths are combined, there is a Oneness of the mission plan. The plan is first to raise one's vibration and consciousness. Once this is accomplished, one should try to raise the consciousness of others. To do this, one must acknowledge the God force within everyone. Many souls are still lost and in a state of forgetfulness. For those it may take many more lives to reach enlightenment.

The starship that was communicating with Norma was called Athena, whose commander was named Juluionno. He told Norma that they "own the Earth and have been here from the beginning of time." At the time, the space-craft was in the Fourth Dimension realm of time/space and was within our realm of understanding and consciousness. Their purpose is to transmit higher consciousness, and they are in charge of the governing body that rules higher consciousness and intelligence that assists humans on their path to higher consciousness.

The Earth has been a training ground for eons, as Arcturians were here long before humans and other space beings. They have bases on the Earth, and many of them are housed inside mountains and other locations. Because they are in the Fourth Dimension, they do not need to fear detection by humans or other potential physical problems. Another base they are connected to is located inside a person's heart. Most of the efforts of the Arcturians go to reawaken humanity through the heart. Arcturian star bases are located all around the universe and in many countries on Earth.

Arcturians are assigned to any group of beings who warrant their protection because of their earned vibration rate and destiny. It has been determined that Earth is worth saving. Some of the Arcturian space-craft are designed for battle, but not many are around Earth today. Because the Arcturians have so many space-craft, the negative forces do not interfere with the planets protected by the Arcturians.

One of the main purposes of the Arcturians is to educate humans on their purpose in life, their powers, their health conditions, and their destiny. They are always working with souls to raise their vibration by reawakening in them knowledge that the soul already knows. Their biggest obstacles are the government and military of many countries.

The Arcturians introduced themselves to government institutions many years ago and were met with hostile and deceitful behavior. Instead of communicating with the government, they now communicate with those in more local control, as these individuals are not afraid to communicate. The Arcturians said they have met with many presidents and premiers. Many of the leaders wanted Arcturian secret technology for the advancement of their country. They have exchanged some information with the military. They told Norma they work with fewer than one percent of the souls in 35 countries. This information is being received and shared, as it is their mission to raise the vibration of the entire Earth.

Their method of contact has been multifaceted. They have provided information in written form, they have appeared physically, and they have taken humans aboard ships with their permission. They also communicate with those souls of alien origin who look predominantly like humans. They rely heavily on these methods of communication. Another method is telepathic, where they send messages and communication to many. They also communicate to souls in their dream state, which they find to be the purest form of communication. During the dream state, each soul has access to the Oneness state.

The Arcturians were asked by the ascended masters, angels, and celestial beings to help Earth move into the Fifth Dimension. They told Norma that it will be another 26,000 years before the planet will have another opportunity to assume the strength and position to go through this experience again. As a result of entering the Fifth Dimension, all structures built by humans for the purpose of separation will come to an end. Only those structures vibrating to Fifth Dimension frequency will replace them.

This new state of consciousness that we will enter deals only with the heart and the amount of love an individual carries. Earth is beginning to prepare for a cleansing of negative energy that surrounds her, as there will be a massive cleansing of Earth. To survive in this new era, individuals must be of a high frequency. The Arcturian mission is to educate the souls of Earth in the way of survival for existence in Fifth Dimensional frequency. Each soul will have to choose its destiny.

OTHER GALACTIC CIVILIZATION MISSIONS

The Arcturians are working with other galactic civilizations to help us get to the Fifth Dimension. Each has its mission during this window of opportunity, which the Arcturians discussed with Norma.

Orion

The Arcturians describe the Orion star system as being very erratic. The Orions have the capability to polarize forces to obtain a balance and beauty that could not otherwise be created. They create havoc and chaos and then try to bring harmony out of chaos.

Orion contributes its mental power for the development of smoothly-running systems on Earth. They are the source of power for many organizational structures of government, business, and energy. They have the main task of influencing the development of networking and linking data systems of the planet.

Lyra

The souls from Lyra will play a major role in Earth's future. They will be sent to lead the discovery of a more advanced process of evolution and harmony which will be more appropriate in Earth's future timetable.

Lyrans are known for their migration qualities and bring forth a freedom of spirit to Earth. Much light is radiated from their head area, but they have difficulty satisfying the heart center while in the Earth's atmosphere. Lyran energy is not used in the lower vibration of Earth. They represent independence of the universe, independence that is required for movement into this new era.

Alpha Centaurus

This advanced civilization, which the Arcturians learned under, has the highest quality of scientific and technical knowledge in the universe. They are theoreticians who are here to raise levels of technical knowledge to the highest realm of theory. One difficulty of the Alpha Centaurians is to make this knowledge understandable to humans. Their information is transmitted through telepathy, but they have such a high vibration, it is difficult for them to ground ideas.

Sirius

The Sirian mission is to help ground the higher forms of information that are coming to Earth. They are here to help us bridge the gaps of theoretical levels of knowledge brought to Earth and the practical application of those ideas. Sirians are the workers and doers. The Arcturians confirmed that the early civilization of Egypt was aided by Sirians who came to Earth at that time and were god-men and god-women. They built the great pyramids and temples with the highest form of knowledge recorded in history. They built pathways and tunnels to the inner world and pathways to the stars allowing them to communicate above and below. They will help build the new shrines and temples in the upcoming Golden Age, assisted by many other star systems. The Arcturians say that the secrets of the Great Pyramid will be uncovered and the mysteries of life will be revealed.

Hydra

Beings from Hydra are excellent in creating with their hands from the substances of the earth. They are the agriculture experts, archeologists, and clay artisans. They have helped transform the Earth's energy into forms of beauty. These sensitive beings love land, labor, and art.

Pleiades

Pleiadians speak the truth and demand justice for all. They come from a star system known for its great advancement in music and dance, and they are here to change the rhythm of the planet with new forms of music, with higher vibrational frequencies that are designed to activate the higher energy vortex within humans. This will help transform consciousness to higher dimensions. Pleaidians are considered to be experts of light and sound creation. Because light and sound comprise all the world's manifestations, the higher frequency tones will open high centers of wisdom. Each galactic civilization has a different mission to help the evolvement of humanity. Extraterrestrials are here now primarily to protect Earth and highlight our spiritual path.

OTHER KNOWLEDGE FROM THE ARCTURIANS

Secrecy

The Arcturians were asked why they have been so secretive. They told Norma that it was not in their original plan. It has been our government and military who kept knowledge of extraterrestrials from all of us. The Arcturians have attempted to contact government officials, but they only want to trade for military technology and information that reveals the mystery of power. The Arcturians want to tell the world about the concept of God and Oneness of the universe. Government leaders are very suspicious of the extraterrestrial agenda of transforming consciousness. They only want scientific knowledge for military purposes. The military has shot down a number of alien space-craft, and this is one of the main reasons they need to be secretive. It is only a matter of time, the Arcturians say, until Earth will be working with extraterrestrials in peace and harmony because they are trying to get us back home in the Fifth Dimension.

Star Children

Star children are born into human form and are here to be a liaison between the enlightened path of their home planet and the enlightened path on Earth. Star children are teaching Earth humans lessons that come naturally on other planets. These children filter higher knowledge. They are of the color violet and know the way to the white light. A common term on Earth is to call them Indigo Children. These children of light are becoming frustrated because their mission is not being revealed to them. The Arcturians say these children, who have assumed the shape of humans, are the finest extraterrestrial life that the planet has known.

Ascended Masters

The Seventh Dimension contains the souls of all those beings who have achieved the vibrational frequency equivalent to the Seventh Ray. They are the Ascended Masters. Once a soul has passed the tests of the seventh initiation on the Earthly plane, the soul is ready to exit the Third Dimensional frequency and journey into the higher levels, the Seventh Ray frequency. Light has seven colors, with violet being the highest frequency band.

The Arcturians are in communication with Sananda, who was Jesus while on Earth. They say the most significant event that will be occurring on Earth will be the second coming of Christ, as it will truly be a reality. This will occur on both the Third and Fourth Dimensions. Jesus is the master of the Sixth Ray and is the head of the awakening process for humanity's new era on Earth.

THE STARSHIP

The Arcturian space-craft contacting Norma was designed to support their mission on Earth, not designed for battle. The Earth mission is to bring enlightenment to humanity. The craft has 35 major divisions described fully in *We the Arcturians*. A few of these parts of the starship will be briefly discussed in this chapter and will help show the advanced knowledge of the Arcturians. The space-craft's name is *Athena*, a symbol of the highest and wisest.

Command Bridge

Athena's command bridge has a vast computer system, even though the Arcturians acknowledge they have long outgrown the use of computers. The computer's power source is crystals which have a means of attracting and conducting light energy from the Great Central Sun. Crystals also power the space-craft which attracts positive and negative forces. This allows the propulsion of the matter/antimatter gravitational pull to flow through various aspects of time and space. The bridge has viewing screens all around it but no windows, as the Arcturians transcend space and time and see with their telepathic abilities.

Reunion Quarters Chamber

Arcturians learned long ago that to survive in a higher realm, they needed to control its population. They learned two primary principles: 1. To control their emotions with thought patterns, and 2. To reproduce is an honor and one of the highest professions on Arcturus.

In the early days of Arcturus, they found it difficult to raise children who would maintain a vibration of the light consciousness. To remedy the problem, they decided upon a selection process for those who were deemed perfect candidates for procreation and to have them ingest energy of the highest caliber.

For births on the space-craft two individuals go into a Reunion Quarters chamber—at the appropriate time to begin the reproduction process. There is no physical contact between them. However, through a mind-link process, these two energies are stabilized in perfect balance. Through the procreation process of the light, the electron force flows through the two beings and conception occurs. Every seven years the reproduction cycle is repeated.

When the new being is created, a celebration occurs. The newborn is taken to a special unit that encases it in the proper vibrational frequency until it is ready to be brought to Arcturus for family unit integration. There are special Arcturians who care for the new life, and those created on starships are brought back to Arcturus. Arcturians are known to seed other planets and have earned the right to be a selected species to perform this function.

Duty Free Port

This area supports a constant exchange of goods between Earth and other worlds where the Arcturians journey. They told Norma that they transport many souls that are destined for embodiment from Arcturus to Earth. They have brought to Earth some of the finest light bearers that the planet has seen. These light bearers are now beginning to wake up and develop the thought process to directly communicate with the mothership.

Remembrance Headquarters

This area of the ship reunites the crew with the home system of Arcturus. It is a replication of the planet itself that has the ability to take the etheric body of the space-craft's crew back to home base.

Recruiting Parcel

The Arcturian mission is to bring enlightenment to humanity, so they must continually leave the ship and manifest on earth. This is the area of the ship where the crew returns and is debriefed. The directory command is located in this area and contains names of earthlings with whom they are in communication. Most of these earthlings are not aware that they are being contacted.

Sleeping Quarters

In this area Arcturians can activate the dream state and make contact with their celestial brothers and sisters. Here, they are also reconnected to their higher existence where they receive spiritual knowledge and guidance. Every seven days they return to the sleeping quarters to access this important aspect of their life.

Chamber for Manifestations

This chamber complements and supports rearrangement of the molecular structure of the Arcturian's central nervous system. It also induces pleasure and performance at the Fifth Dimension and higher. The Arcturians emphasize that they are here to help Earth humans to transform into the Fifth Dimension with as little pain as possible.

Learning Facility

This area is designed for enhancement of knowledge, including a data bank of material regarding every aspect of Earth. Arcturians learn by ingesting information through their telepathic abilities and central nervous system. Arcturians can absorb concepts and information 100 times faster than a human.

Locomotion Chamber

This chamber is designed to facilitate movement while in the Earth's atmosphere and protect Arcturians from outside forces that could hold them in Third Dimensional existence. The energy of Earth is out of balance to the masculine side. All beings who enter the ship are required to use this chamber which balances this energy. A balanced being has the ability to raise or lower his or her vibrational frequency to any desired level.

Chamber for Captives

Great souls have come to earth to open energy vortices to be accessed in the future. These souls are embodied in a sea of negativity when on Earth, which makes it difficult to see through the illusion, especially those isolated on the planet. Arcturians take these souls aboard to help remove the energy blockage. The chamber assists souls to reawaken. Their mission is to warn us that time is running out and we are at a point in time that will determine our fate.

Communication Chamber

This chamber supports growth and development of telepathic power. The ship has an instrument that absorbs mental thought and records their significance and power. The readout helps Arcturians determine their own power and potential for their telepathic abilities and lets them know what areas need improvement. It is used for both learning and skill development for telepathy. Only accurate information is transmitted and received. Each being is required to use the chamber once a month.

Navigation Rehearsal Area

Much like a flight simulator for airlines, Arcturians can practice a combination of maneuvers with their starship. They have a sensing computer that reacts to energy movement such as thoughts, physical forms, and even emotion. The starship is one of the few with this chamber that many visiting extraterrestrials can take advantage of.

Motion Chamber

This area performs an integration of necessary information and data into a holistic design, and programs the holistic data into the essence of a being who journeys outside the ship. Those leaving the ship are led into the chamber which reprograms the soul to ensure the being receives all the information necessary to perform the next part of the mission.

Shuttle Craft Area

A Shuttle craft's purpose is to access energy points on Earth and to reactivate them. They are global in shape and small in size. This keeps Earth in alignment with its electromagnetic current. For example, they have to correct for the currents emitted by nuclear power stations that affect the natural alignment around the planet. These shuttle craft also transport some humans to the starship. The Arcturians gave humans their current knowledge about aerodynamics. These ideas were transmitted from a higher collective consciousness.

Information Chamber

The information contained within this chamber contains the propulsion matter/antimatter formulas necessary for advancement of hyperspace travel. Arcturians guard this information carefully, especially from any malevolent extraterrestrials. This area contains the most powerful information of the universe.

The Motion Carrier Chamber

Much of what was carried in the cargo holds of the starship has been deposited at the Arcturian bases on Earth, mainly inside mountains within the crust of Earth. They also have three bases on the moon that are of the same dimensional frequency as fairies and little people.

The cargo has to be in a pure refined state to exist in the middle of matter, such as mountains. The walls of the ship are designed to hold the frequency of each piece of cargo. This area of the craft is most vulnerable to attack. However, most malevolent universes stay away from the Arcturian ships because of their reputation for being a high-minded civilization with advanced technology.

Intelligence and Debriefing Chamber

This chamber holds the essence of Arcturian Earth missions and is used by both Arcturian officers and beings from other races. The purpose is to exchange information here. Souls are often brought here in their dream state to work with the Arcturians for Earth projects. After the discussion is over, the souls are returned to their physical body. Many Earth souls are being prepared to work with the Arcturians in the higher realms. They freely come aboard after an awakening process and participate in Arcturian plans for raising the frequencies on Earth.

Intelligence and Vaporizing Chamber

In this chamber the Arcturians manifest forms of gifts that are products of Arcturian thought forms. Arcturians use this area for manifesting and vaporizing ideas and objects. They consider themselves to be very giving people, believing in the law "The more you give, the more you receive." Some of this technology has been given to scientists on Earth.

Light Rejuvenation Center

This area produces the power that propels the starship. It transforms the energy from crystals that produce power into liquid light. Liquid light is drawn through the crystal from the universe and is locked into the transformational process. This is the life support system of the ship.

Mechanism Chamber

This chamber allows access to many places outside the normal areas of operation. It transforms physical matter into another dimensional frequency, recording the exact molecular structure which contains the program, through hyperspace and reforms it in another location. The chamber holds the program for individual life forms. If a malfunction occurs, the being can still be changed back into its original form. This chamber also decodes a life form's genetic structure. It can also detect defects in a being's health and reprogram it back to health, serving as a medical facility. There is no illness on an Arcturian starship.

Engineering Apparatus Area

This area can maneuver the craft between the Third, Fourth, and Fifth Dimension and is designed to cause abrupt transition of magnetic frequency that might be necessary for the ship to become visible or invisible to the human eye.

Engineering Crystalization Area

The silicon found in crystals is a powerful communication device, so crystals are used to make contact with power points on Earth when there is an ongoing need to energize many locations on Earth. The Earth is constantly changing its influx of magnetic energy impulses and has a tendency to shift polar point connections that are needed for communication.

The Arcturians also use crystal energy to raise consciousness in humans. The Earth receives this information through vortex areas of electromagnetic energy which is distributed throughout the planet by ley lines and grid lines. Norma was told that human consciousness needs to adjust slowly to the higher frequency the new era will demand.

Arcturians often implant gigantic crystal beds in select locations on Earth. Much UFO activity occurs in these areas because they are communication bases. If the Arcturians could not produce and program the crystalline structures as they do, humans would not be as far along in consciousness.

Influx Setting

The Arcturians told Norma that any system of command, such as the military, that does not allow an individual to express his free will does not support the God concept. God made humans in the likeness of God with the power to cocreate the universe. One of humanity's lessons is to allow another person his freedom. If a person blindly follows another person's will, that soul will jeopardize its own karmic debt.

The influx setting on a starship is a monitoring instrument that allows each Being to calculate the vibration frequency it has achieved. Because the Arcturians have mastered their emotions, they can use this device to aid in their evolution. The results are shared with the Elders who counsel the beings on their life journey. Because Arcturians are in a foreign environment on their missions, they need feedback regarding their well being, which is why they have the instrument on board.

The Arcturians have been hard at work to raise humanity's vibration frequency. The window of time is small to get us to this higher frequency for the Fifth Dimension. We have come a long way, but still have a way to go to practice forgiveness and love and understand the concept of Oneness. Once humanity has manifested a certain vibratory frequency, we will understand Oneness. Our friends the Arcturians think we will make it and will continue to work with us.

Chapter Seven

THE ANUNNAKI
Earth's Ancient Gods

Ancient history tells of the important role gods have played in early civilization. Sumer, Babylonia, Egypt, Greece, Rome and many more civilizations have had a mythology of gods. The Bible refers to gods in Genesis, and Plato writes of gods being rulers of Atlantis. These gods have remained a mystery to scholars, with most believing they were a fabrication by man to help understand unexplainable events. It has been the research of Zecharia Sitchin in recent times that has given reality to the gods. Sitchin is a multi-lingual, historical researcher who is one of the few people with a command of the ancient Sumerian cuneiform language, as well as Hebrew and several other ancient languages. This chapter is based on Sitchin's research that he has recorded in seven books.

One of the greatest mysteries of ancient history is explaining how the world's first recorded civilization advanced so rapidly in the land of Sumer. Before the middle of the 19th Century, historians were unaware of this ancient culture. Sumer comprised the lower half of Mesopotamia, a land today occupied by modern Iraq from north of Baghdad to the Persian Gulf, an area of approximately 10,000 square miles. Two major rivers, the Tigris and Euphrates, provided its lifeblood. The records of this ancient civilization were preserved on clay cuneiform tablets that provided a deep insight to this earliest recorded civilization and to the gods who started it.

In classical times Sumer became known as Babylonia. The climate was hot and dry, the soil arid and windswept. No trees provided timber resources, and large reeds covered many marshes. Beginning in the fourth millennium B.C., the

133

ingenuity of the Sumerians transformed this forsaken land into a Garden of Eden.

Sumer was the land of firsts. It provided recorded history with the first written documents, the first schools, the first political congress, the first recorded history, the first taxes, the first legal code, and man's first cosmogony and cosmology. As one compares this ancient civilization with today's modern society, not many things have changed over the millennia. Historians believe Sumer laid the foundation for our current civilization.

Scholars are bewildered how a great civilization could arise from nowhere in such a short period of time. They have no idea where the people of Sumer originated. Legends that explain the genesis of civilization in Mesopotamia infer there was an influx of people from the sea. One must wonder if Sumer inherited this great civilization from Atlantis. Sitchin's research into the ancient Sumerian cuneiform writings strongly suggests that civilization on Earth was seeded by extraterrestrials, the Anunnaki.

Sumerian texts, according to Sitchin, tell about the DIN.GIRs, who are "the righteous ones of the rocket ships." They had come to Earth from their own planet, and chose the land between the Tigris and Euphrates as their home away from home. They called it KI.EN.GIR, "the land of the lord of the rockets." The Sumerian texts date the Anunnaki's arrival at 432,000 years before the Deluge when the DIN.GIR came to Earth from their own planet. The planet's name was Nibiru, which Sumerians considered the twelfth member of the solar system. As we will see, Sitchin's Sumerian research has provided us answers about the creation of our planet, of man, of the ancient gods, and of the mysteries of the Bible.

THE CREATION OF EARTH

Sumerian texts tell of our solar system consisting of the sun and eleven planets, counting the moon. Nibiru, the twelfth planet, was the planet of the DIN.GIRs, whom Sitchin calls the Nefilim. Nefilim is a term used in Genesis, traditionally thought to mean giants. However, the literal translation in Hebrew means those who were cast down. Chapter 6:4 of Genesis reads,

"There were giants (Nefilim); and also after that, when sons of God came unto the daughters of men, and they bore children to them . . ." The sons of God were the Nefilim. The Biblical writings parallel the Sumerian texts that tell of the Nefilim cohabiting with the daughters of men and bearing children.

The creation of Earth was before the creation of man. The writings of Sumer tell of an unstable solar system, consisting of a sun and nine planets invaded by a large comet-like planet from outer space. This invading planet was Nibiru, which the Babylonians called Marduk, and it had seven satellites. It was on a collision course with a large planet named Tiamet, a planet with water that was found between Jupiter and Mars. The two planets did not collide, but one of Nibiru's satellites smashed into Tiamet and left it fissured and lifeless. Nibiru's passage brought it into the gravitational pull of the sun causing it to hit Tiamet on its next orbit around the sun, resulting in Tiamet splitting in two. Another of Nibiru's satellites hit one of the separated halves of Tiamet and brought it to a new orbit where no planet had orbited before, and this became Earth. The other half had a different fate; on the next orbit of Nibiru, it collided with the remaining half of Tiamet and left it in many pieces, resulting in the asteroid belt the ancients called the "Great Band" or "Bracelet." This collision also caused Tiamet's chief satellite, Kingfu, to fall into the gravitational pull of Earth and it became the moon and the remaining pieces of Tiamet became comets. This explains the concentration of Earth's continents on one side and a deep cavity on the other side, the Pacific Ocean. They called Tiamet the "water monster," which explains where Earth got its water.

Because of the collisions, Nibiru was then bound to the gravitational pull of the sun. The direction of its elliptical orbit was opposite that of the other planets. A 4,500-year-old Sumerian depiction of the twelve celestial bodies shows the twelfth planet orbiting between Mars and Jupiter with an orbit taking 3,600 years.

Astronomers have long been looking for evidence that another planet existed between Mars and Jupiter. Historically, all the planets beyond Saturn were discovered mathematically, not visually. Mathematical calculations, called Bode's law, predicted that a planet should exist between Mars and Jupiter. Three thousand asteroids have been discovered, believed to be debris that scientists cannot explain from a planet shattered to pieces.

THE COLONIZATION OF EARTH

The planet Nibiru contained an abundance of radioactive elements that generated its own heat from within. Volcanic activity provided Nibiru with an atmosphere that ancient texts describe as a halo that clothed Nibiru, which enabled the planet to retain its own heat. The Sumerian writings do not tell if Nibiru was colonized after its collision with Tiamet or if there were survivors. At any rate, the atmosphere began to wane, so the shield that trapped the heat began to dissipate. The only solution to prevent dissipation was to suspend gold particles in the atmosphere. This was so crucial that a space program was developed whose sole purpose was to provide gold to the twelfth planet.

Anu was the ruler of Nibiru and chose his son Ea, later known as Enki, a brilliant scientist and engineer, to be the leader of the first mission to Earth. All the ancient texts agree that it was the god Enki, accompanied by fifty Anunnaki, who waded ashore to the edge of the marshland and said, "Here we settle." These first colonizers arrived 432,000 years before the Flood, with the original plan to extract gold from the waters of the Persian Gulf. They named the first settlement Eridu, in the land of Sumer.

The colonists worked hard once they arrived, dredging beds of streams, filling marshes, and building dikes. Enki spoke of much flooding when he first arrived on Earth as this was during an ice age. Later, the ice had begun to melt, accompanied by much rain. Enki enjoyed the water and later was assigned as the god in charge of the watery deep. He purified the Tigris River and preferred to travel by boat. Enki and the first group of Nefilim remained on Earth for eight Shar, or 28,800 years. One Nibiru year equaled 3,600 Earth years, the time it took for one orbit of Nibiru to orbit the sun. While Enki was enduring hardships on Earth, his father Anu and brother Enlil were watching the developments from Nibiru. It was Enlil who was really in charge of the Earth mission.

Enlil came to Earth to take personal charge of the mission. He settled in Larsa, also located in Mesopotamia, and stayed there for six Shar (21,600 years). A Mission Control Center was established nearby at Nippur. An epic poem tells of Nippur being protected by awesome weapons. An artificially raised platform was found at the Mission Control Center that was the communication center of Mission Control, the bond between "Heaven and Earth." Once established, it also became a supply depot for abundant supplies brought by shuttle craft from Niburu. Sippar was the spaceport of the Nefilim located northeast of Nippur.

Sumerian seals depict boxlike divine objects transported by boat and pack animals. Sitchin describes the depictions as similar to the Ark of the Covenant built by Moses with directions from God. Implications now suggest that the Ark of the Covenant was a communication box electrically operated, and no one was to touch it or he would die immediately by electrocution. To prevent electrocution, the Ark was carried by means of wooden staffs passed through four golden rings.The Ark enabled communication with a deity.

A winged globe symbolized the planet Nibiru in the ancient writings. Wherever archaeologists discovered remains of Near Eastern civilizations, the winged symbol was found. It was dominant on temples, palaces, king's thrones, chariots, and cylinder seals. The rulers of Sumer, Akkad, Babylonia, and Assyria all revered the symbol.

When Nibiru approached Earth on its 3,600-year orbit, it signaled upheaval, great change, and the beginning of a new era. It was a predictable event. The Great Flood occurred when Nibiru made its nearest orbit to Earth. The planet could be seen from Earth in the daytime, and Sumerians referred to it as the Day of the Lord. The most important religious event of Sumer was the twelve-day New Year's Festival celebrating the orbit of Nibiru that coincided with the spring equinox.

The original plan of the gods was to acquire gold from the sea, but they soon realized that it was not feasible. They discovered gold deposits in the mountains of southeastern Africa near grassy plains and lush vegetation. The main deposit was found at Mount Arali where mining soon commenced, and the land became known as Abzu. The Sumerian pictograph of Abzu was that of an excavation deep into the ground mounted by a shaft. Working hard, the Anunnaki (Nefilim) did the mining themselves. Special cargo ships carried the gold back to Mesopotamia where the gold was cast into ingots. Enki's son Gibil was in charge of smelting the gold to ready it for transportation back to the space station where the Annunaki transferred it to spaceships en route to Nibiru. The twin peaks of Mount Ararat were used as a landmark for the space-craft, and they laid out all the Sumerian cities to form an arrow pointing to the spaceport at Sippar. Enki became a frequent commuter between Sumer and Abzu, overseeing the gold project. Scientific studies have shown that mining had occurred in Swaziland (South Africa) about 70,000 - 80,000 B.C. The cuneiform texts suggest the Anunnaki were also mining radioactive material such as uranium or cobalt. Pictographs show powerful rays emitting from a mine and the gods attending Enki using a screening shield.

The Sumerian writings describe the Anunnaki, numbering 600, as the rank and file gods who settled the Earth, and these few gods performed the physical labor. Another 300 Anunnaki astronauts, called the Igigi, remained in the space-craft providing support.

The Anunnaki toiled hard and long. Digging was the most common and arduous chore, which they all abhorred. The lesser gods dug the river beds to make them navigable, canals for irrigation, and mine shafts in Abzu to bring up the minerals. They worked for forty Shar, or 144,000 years, and finally reached a point where they could not tolerate it anymore and cried out, "No more!"

THE CREATION OF MAN

When the Anunnaki settled Earth, man was not yet here, according to the Sumerian writings. Man originated because of a mutiny committed by the Anunnaki who refused to work, at a time when Enlil was visiting. It upset Enlil so much that he wanted to resign, especially when Anu sided with the Anunnki. It was Enki who offered a solution to the crisis, suggesting that a primitive worker be created to take over the manual labors of the Anunnaki. It was a very popular suggestion with the ruling gods who voted unanimously to create such a worker. His name was to be man.

It was Enki and his half-sister Ninhursag, the Mother Goddesss, who were given the task to create the primitive worker man. The great experiment took place in a laboratory and was based on genetic engineering. Scholars have long been searching for the missing link in man's evolution, which Sitchin believes was genetic manipulation in the laboratory 300,000 years ago by the Anunnaki.

After much experimentation, Ninhursag purified the essence of the sperm of young male Anunnaki. She then mixed it into the egg of a female ape. Once the egg became fertilized, they implanted it into the womb of a female Anunnaki for the remainder of the pregnancy. When the hybrid creature was born, she lifted him up and shouted, "I have created it. My hands have made it." It had taken the Anunnakis' considerable trial and error to achieve the perfect model. Fourteen Anunnaki birth-goddesses were implanted with the genetically manipulated egg of an ape woman. The new being was called the "Adam" because he was created of the Adama, the Earth soil. Man was similar to the gods, both physically and emotionally, because they created him that way.

As the hybrid babies grew up, they put them to work in the mines. Originally they sent the slave workers to Abzu to work the mines. As a result, they mined more ore in Africa, resulting in a greater workload in the processing facility in Sumer. The Anunnaki workers in Sumer began to clamor for slave workers, but Enki initially refused. Enlil, wanting the additional workers, attacked the African mines, resulting in an agreement that the workers labor in both lands. In Sumer, the genetically engineered slaves were used for processing the ores, digging canals, and raising food.

Sitchin believes the Sumerian writings have given validity to both the theory of evolution and to the Biblical story of creation, which also originates from Sumer. It was a deliberate creation of the gods and explains the gap in evolution. They created man to do the work of the gods and be their servants. The decision to create man was made by an assembly of gods. In the Book of Genesis, the Elohim said, "Let us make man in our image, after our likeness". Elohim is the plural of deity.

Anthropologists have scientific evidence that man evolved and emerged in southeast Africa. The Sumer texts also suggest the creation of man took place in Abzu, in the Lower World, where the mines were located in present day Africa. Science has determined that Homo Sapiens inexplicably appeared some 300,000 years ago. The Nefilim (Anunnaki) landed 450,000 years ago, toiled for 144,000 years, and created man 300,000 years ago. The Nefilim, under the direction of Enki, took an existing creature, the Homo Erectus, and manipulated it in the image of the gods by means of genetic engineering. What was needed next was a quick process for mass production of new workers. They continued to experiment. Non-Sumerian texts tell of hideous beings being created. Some men had two wings, some had two faces, some had goat horns, and some had horses' hooves. The Anunnaki tried to hybridize the new creature with other animals.

Enki and Ninhursag in their experimentation created six deformed humans. Once they achieved the perfect man, named Adapa, the Mother Goddess gave man the skin of gods. In the final product, the Anunnaki were genetically compatible with the daughters of man. They were able to marry them and have children with them. This was only possible if man had developed from the same seed of life as the Anunnaki, referred to as Nefilim in the Bible. Man was a mixture of a godly element and the clay of the earth.

Anu referred to Adapa as the human offspring of Enki. It was Ninki, Enki's wife, who was pregnant with the first Adapa. It was the perfect mold and gods clamored for more. Duplicates were either male or female, and Eve was made from Adam's essence. The divine element came from the male genes, and the earthly element came from the female genes.

Enlil wanted to keep the humans sexually suppressed, while Enki, on the other hand, wanted to bestow upon mankind the "fruits of knowing," procreation. It was Enki who genetically engineered man so he was able to have children. This unauthorized deed angered the gods, and they arrested Enki. The enraged Enlil ordered the expulsion of the Adam, the homo sapiens earthlings, from Edin the "abode of the righteous ones." Man was no longer confined to the settlements of the Anunnaki, and man began to roam the world.

EARLY MAN

Ancient texts tell of the god's blood being mixed into the clay so as to bind god and man genetically to the end of days. Both the image (flesh) and likeness (soul) of the gods would be imprinted upon man in a kinship of blood that could never be severed. The Akkadian term for clay is *tit*, referring to the ovum of a female ape (homo erectus) fertilized by the genes of a god's sperm.

As long as Adam and Eve lacked knowledge about sex, they were allowed to live in the Garden of Eden. The Sumerian name for the god's abode was Edin, the home of the righteous ones. Enki made the decision to give man the ability to procreate, angering Enlil. Sitchin suggests that the metaphor of the serpent in the Bible represents Enki who gave Adam and Eve the knowledge of procreation. The deity represented Enlil who cast Adam and Eve out of Edin, the abode of the gods. Knowledge of sex for procreation was a crucial step in man's creation.

At some point, the banished humans were allowed to return to Mesopotamia to live alongside the gods to serve and worship them. In the Bible during the days of Enosh, "the gods allowed mankind back into Mesopotamia to serve the gods." Enki took Adapa (Adam) by space-craft to see Anu in Nibiru. After his return, man proliferated. Men were no longer just slaves in the mines and fields, but they performed all tasks. They built houses and temples for all the gods and learned how to cook, dance, and play music for the gods.

In the fourth generation after Enosh, according to Genesis, the firstborn son was named Enoch. He walked with the deity, and according to the Old Testament, he did not die on Earth, but was taken by the deity. The Book of Enoch expounds on this and details his first visit with the angels of God to be instructed in various sciences and ethics. After returning to Earth to pass on this knowledge and the requisites of priesthood, he was taken up permanently to join the Nefilim in their celestial abode.

Young Anunnaki were lacking female companionship and began to have sex with the daughters of man. The Sumerian writings parallel those of Genesis. In those days when the Nefilim (Anunnaki) were upon the Earth, they cohabited with daughters of the Adam, and they bore children with them. They were mighty ones of eternity, the people of the shem (rocket). The good life was the main concern of the Anunnaki, and mankind became too infatuated with sex and lust. Genesis describes the taking of wives by the Nefilim. After awhile, the genetic perfection of mankind began to deteriorate. What upset Enlil to no end were the increased sexual relations between the male Anunnakis and the daughters of man. He cried, "Enough!"

The Bible reflects this concern by Enlil, and the Lord said, "I will destroy the Earthlings whom I have created off the face of the Earth." Enlil saw his chance to eliminate man when a scientific station at the tip of Africa reported an ensuing natural calamity. Sitchin believes they saw that the growing ice cap over Antarctica had become unstable and was resting on a layer of slippery mush. Because Nibiru was making a close pass, the gravitational pull would cause the ice caps to slip into the ocean resulting in mass flooding.

The Deluge was a predictable event but an unavoidable one. The gods conspired not to tell mankind of the upcoming disaster, as this was their chance to eliminate mankind from Earth. Before this opportunity of elimination, Enlil had called for plans of natural disasters to decimate mankind through pestilence, sickness, drought, and starvation. But now there was even a better plan as offered by Enki who told the assembly of ruling gods about the upcoming flood. The gods swore Enki to secrecy, forbidding him to tell man of his upcoming demise.

Enki was determined to save his creation, mankind, but he also was obligated to follow the wishes of the gods. He instructed his Noah, King Ziusudra of the city Shuruppak, in the temple where he hid behind a screen. Enki whispered urgent instructions to Ziusudra to construct a submersible boat, a submarine, that could withstand an avalanche of water. Precise dimensions were given. He called for a boat roofed over and sealed with tough pitch. There were to be no decks or openings. Enki provided Ziusudra with a navigator who was to direct the vessel to Mount Ararat, the Mount of Salvation. The sons of Ziusudra and their families were also to board the vessel.

When the storms that preceded the Deluge commenced, the Nefilim boarded the shuttle craft and orbited the Earth until the waters subsided. As the fleeing gods watched the destruction of Earth from above, they trembled from the noise caused by the Deluge and then realized how much they had fallen in love with the planet. Ninhursag was said to have wept with other Anunnaki as they were all humbled.

When the water subsided, Enlil saw the ark on Mount Ararat and was filled with anger because there were human survivors. Only after Enki convinced Enlil to make peace with the remnants of mankind did Enlil accept the fact that mankind would survive.

The families of Ziusudra were sent out to settle the mountain ranges flanking the inundated plains of the Tigris and Euphrates Rivers. They were to wait until the plains were dry enough to cultivate before they inhabited them. Cereals and grains were brought down from the mountains and planted under the direction of the god Ninurta. Before the Deluge, seed had been taken to Nibiru and was reintroduced to Earth. Ninurta also dammed the mountains and drained the plains. Genesis describes sowing and harvesting as Noah's divine gifts. Scholars have placed the origins of agriculture about 13,000 years ago in the mountain terrain about the time of the Deluge, but they can't explain the earliest grains and domestication of animals. It was Enki who introduced domestic herds. The plan was introduced, and life began to normalize.

Enki returned to Africa to assess the damage. He turned his attention to the Valley of the Nile and utilized his knowledge to recover this great area. The Egyptians have acknowledged that their great gods came from the "olden place."

The Nefilim decided it was time to establish an intermediary between themselves and mankind. Before the decision was made, there was no kingship in the land, as the rules came from the gods. Kingship was decided upon after Anu made a visit to the Council of the Great Gods, who thought the Anunnaki were too lofty for mankind.

The gods decided that Enki must share the divine formula of civilization with the other gods so they could reestablish their urban centers. Civilization was to be granted to all of Sumer. They then made the decision to establish new cities alongside the older ones that had been destroyed in the Deluge. The first city established was Kish, which was put under the control of Ninurta, the son of Enlil, and it became Sumer's first administrative capital. The urban center of Ur was next established under the control of Nannar/Sin, the first-born son of Enlil, becoming Sumer's economic capital. The men appointed kings of the cities by Enlil were called Lugal. The Old Testament confirms the Sumerian writings in Genesis, chapter 10:

Kish begot Nimrod
He was first to be a mighty man in the land
And the beginning of his kingship
Babel and Erech and Akkad
All in the land of Shin'ar (Sumer).

All mankind at one time spoke one language. Initiated by Enlil, the gods decided to divide the tongue and disperse mankind as told in the Sumerian texts and the Biblical story about the Tower of Babel (Hebrew for Babylonia). The original Akkadian definition of Babelli meant "gateway of the gods," the place where the gods were to enter and leave Sumer. Sometime between 3450 and 3600 B.C., the Bible tells of a group of perpetrators who planned to construct a tower whose top reached into the heavens. These ancient words describe a ziggurat. After the people settled in Sumer, they learned the art of brickmaking; they built cities, and raised high towers. They planned to make for themselves a shem (rocket) and a tower to launch it. According to Sitchin, the Sumerian meaning of "mu" and its Semitic derivative "shem" refers to the heavenly journeys of the gods by sky vehicles. Sumerian pictographs verify multistage rockets and flying vehicles for roaming the Earth's skies. Unearthed artifacts show pictures of rockets with wings or fins, reached by a ladder. Sculptures have shown a god to be inside a rocket-shaped chamber. Both the Bible and Sumerian writings imply the flying machines were meant for the gods and not mankind. Man could only ascend to the heavenly abode upon the invitation of a god. The gods became quite concerned about this action by man and feared the human race would unify in culture and purpose, thereby reducing their power and control over man. For this reason, they thought they could control man better with the philosophy of divide and rule. As a result, man was dispersed and given a divided tongue of diverse languages.

Several major civilizations were seeded from this decision of the gods. About 3200 B.C. kingship and civilization made its first appearance in the Nile Valley. Another advanced civilization occurred in the Indus Valley of India that included large cities, a developed agriculture, and flourishing trade about 1,000 years after the beginning of the Sumerian civilization. Scholars believe that the seeds of these two civilizations originated from Mesopotamia, confirmed in the Bible. Following the Deluge, mankind was divided into three branches, led by the three sons of Noah. The lands of Shem inhabited Mesopotamia and the Near Eastern lands. Ham inhabited Africa and parts of Arabia, while the third son Japeth settled the Indo-European lands of Asia Minor, Iran, India, and Europe. Each of these lands was assigned to one of the leading deities. The Sumerian language was differentiated into three different languages following the Tower of Babel incident – Sumerian, Egyptian, and Indo-European.

Following the Deluge, the Nefilim held lengthy council meetings regarding the future of gods and man on Earth. They decided to create four regions including Mesopotamia, the Nile Valley, and Indus Valley. The fourth region was dedicated to the gods alone. Trespassing by men could lead to their quick death. This land was called Tilmun (the place of missiles) in the Sinai. A new spaceport was established here after Sippar was destroyed in the Flood. Tilmun was under the command of Utu/Shamash, grandson of Enlil and father of Inanna, the god in charge of fiery rockets. Many Sumerian tales show man trying to reach this forbidden land with the belief they would acquire immortality among the gods of heaven and Earth.

To end the feud between the Enlil and Enki families, lots were drawn among the gods to determine what region the gods were to have dominion over. Asia and Europe were assigned to Enlil and family, while Africa was given to Enki. Ninuta, Enlil's son, was assigned the lands of Elam, Persia, and Assyria. Nanna, another of Enlil's son, was given the land of Sumer. He was a benevolent god who supervised the reconstruction of Sumer and rebuilt the pre-diluvian cities at their original sites. New cities were also constructed. The youngest son of Enlil was Ishkur (Adad) who was given the northwest lands of Asia Minor and the Mediterranean islands that spread to Greece. Like Zeus, he was depicted as riding a bull and holding forked lighting.

Enki's son Nergal was given southern Africa. Gibil, another son, was given control of the African gold mines. Marduk, Enki's favorite son, was taught all the sciences and astronomy by his father. According to Sitchin, his Egyptian name was Ra, who presided over the Nile Valley. Later, around 2000 B.C., Marduk usurped the Lordship of Earth from Enlil and was declared supreme god of Babylonia.

It was only in the 20th Century that the third region was unearthed in the Indus Valley, and two major cities were found, dominated by an acropolis. The Sumerian goddess Inanna whom the Akkadians knew as Ishtar ruled this area. This region lasted only one millennium, and by 1600 B.C. the civilization no longer existed.

A NUCLEAR CATASTROPHE

Sitchin presents evidence that the Biblical story regarding the destruction of Sodom and Gomorrah was in all likelihood a story about a nuclear holocaust. The Biblical story begins with Abraham camped near Hebron when the Lord disclosed to him the true purpose of his journey. He was to verify the accusations made against the cities of Sodom and Gomorrah. The Lord disclosed to Abraham the upcoming destruction of the two cities, and told Abraham that if there were 50 righteous ones in the city, He would call off the holocaust. Abraham bargained with the Lord to get the number down to ten, and the Lord agreed.

Two visiting emissaries of the Lord were visiting Lot in Sodom when an unruly crowd gathered outside of Lot's home and tried to break their way in. The emissaries persuaded Lot and his wife to leave the city without delay and escape to the mountains without looking back.

The weapon of the gods was discharged, and all the people, cities, and plant life were destroyed by the tremendous heat of the weapon. Lots' wife looked back and was turned into a "pillar of vapor." The Bible says it was a "pillar of salt," but Sitchin asserts the correct Hebrew interpretation is a "pillar of vapor."

Abraham got up early in the morning after the blast and viewed Sodom and Gomorrah from 50 miles away. He saw smoke arising from the Earth and that the once populated valley with five cities was now a new southern part of the Dead Sea. Today it is still known as "Lot's Sea."

Archaeological evidence shows that the region was abruptly abandoned in the 21st Century B.C. and was not reoccupied for several centuries. The water that surrounds the Dead Sea is still contaminated with radioactivity, enough to cause sterility. The year of the holocaust was 2024 B.C., the sixth year of the reign of Ibbi-Sin, the last king of Ur.

The Sumerian texts give a background on why this catastrophe happened. During the sixth year of Ibbi-Sin's reign, Marduk returned to Babylon for the second time as predicted. For 24 years Marduk had been in exile among the Hittites. He wished to bring peace and prosperity back to the land. To ensure Marduk's return, Amorite supporters of Marduk swooped down the Euphrates Valley toward Nippur. The god Ninurta influenced the rulers to organize Elamite troops to fight the invading force, and the two armies met each other at Nippur. Ekur, the shrine to Enlil, was destroyed in battle. Ninurta falsely accused Marduk of desecrating Enlil's holy of holies at Nippur, but in actuality it was Ninurta's ally Nergal that caused the destruction. Enlil became so enraged that he took sides against Marduk and his son Nabu. A plan was made to destroy the centers of support for Marduk at the Canaan cities on the Jordan plain, the location of Sodom and Gomorrah.

Meanwhile, at the Council of Gods, Nergal recommended the use of force against Marduk. Enki lost his patience at the meeting and shouted at Nergal to get out of his presence. Nergal had become aware of the location of seven awesome weapons that he planned to use created by Anu and hidden underground in the mountains. Gibil warned Marduk about Nergal's scheme, and Marduk rushed to his father Enki, who did not know where the weapons were stored but that they would make the land desolate and cause the people to perish.

Ninurta returned to the Lower World (Earth) and found that Nergal had already primed the seven awesome weapons. They warned Nergal that the weapons could only be used against specific targets approved by the gods. Mankind had to be spared. Nergal agreed to notify the Anunnaki and the Igigi manning the space station, but he refused to warn Marduk. He used words identical to those attributed in the Bible to Abraham when he tried to have Sodom spared.

Nergal's strategy was to deny Marduk his greatest prize, the spaceport located on the Sinai. When Nergal presented the plan to the gods, Ninurta was speechless, but Anu and Enlil approved the plan. The first target was the spaceport, and it was Ninurta who detonated the first nuclear bomb. Nergal is the god who decimated Sodom and Gomorrah.

The Anunnki guarding the spaceport in the Sinai were warned so they could escape to the orbiting space station. The spaceport was obliterated, the runways demolished, and all vegetation decimated. Today there is a black scar on the face of the Sinai where the explosion blackened the soil.

The explosion had a profound effect on Sumer that essentially caused its demise. The blast gave rise to a great radioactive wind that blew from west to east. Sumer became the ultimate victim. Unearthed lamentation texts describe the ordeal in Ur, Nippur, Uruk, and Eridu, which experienced a sudden concurrent catastrophe. Cities were without people, the pastures were without cattle and sheep, the rivers were bitter, the fields turned to weed, and the plants withered. Resident gods abandoned the cities.

The source of this unseen death was a brownish cloud with a luminous edge that appeared in the skies of Sumer at night.

A fast howling wind, described as evil, accompanied the cloud caused by a huge explosion from the awesome weapons. The unseen death episode lasted 24 hours.

The gods were upset; Ninhursag wept bitter tears and Nanshe cried. The texts describe Inanna departing from Uruk to Africa in a submarine complaining she had to leave her jewelry. The resident deities had instructed the people of Uruk to run away and hide in the mountains. Enki left Eridu, and he too wept with bitterness. He led the displaced survivors to the desert with instructions to hide underground. Ningal and Nannar spent a nightmarish night in the Ziggurat at Ur. Upon departing the city they saw death and desolation with dead bodies that had "melted away." All of southern Mesopotamia was incapacitated, as its soil and water were poisoned, its cities desolated, and the reed marshes rotted. The end of a great civilization had come to pass.

THE INTERACTIONS OF THE GODS

The Sumerian tablets reveal that the relationship among the gods make modern day soap operas look rather tame. Gods were allowed any number of wives and concubines, but a certain protocol had to be followed. Lovemaking between brother and sister was allowed, but marriage was prohibited. However, marriage between a god and his half-sister was condoned. The birthright of an offspring stemmed from a code of sexual behavior based not on morality but sexual protocol. A first male offspring from a relationship between a god and his half-sister had priority over the firstborn son of a god. The son of a god and his half-sister was believed to have a 50 percent more pure seed. If gods were married to more than one person, they had to select one as their official spouse, preferring a half-sister for the role. The god's official spouse was honored with a feminine aspect of his title. Anu's wife was named Antu, while Sud, the nurse who married Enlil, was named Ninlil. They were married because Enlil raped Sud on the first date and impregnated her. Enki's wife Damkina was known as Ninki.

Enki was the firstborn son of Anu whose mother was Id, one of six concubines of Anu. Anu's half-sister Antum bore him

Enlil. Anu had 80 children, 66 by his other concubines and 14 by Antu. By the Nibirun code of succession, Enlil became the legal heir to Anu instead of Enki. A strong rivalry and jealousy developed between the two half-brothers. When Anu decided to bring Enlil to Earth, heated arguments arose from Enki.

Ninhursag was a daughter of Anu, but not of Antu. She was a half-sister of Enlil and Enki. Ninhursag was one of the original Great Anunnaki known as the Mother Goddess and a member of the Pantheon of Twelve. Enki desired to have a son with Ninhursag but she bore him a daughter. Frustrated, Enki mated with his daughter who also gave birth to a daughter. Not willing to give up, Enki made love to his granddaughter who presented him with another daughter. This incensed Ninhursag who put a curse on Enki that the gods later made her remove. Meanwhile, his brother Enlil had better luck with Ninhursag, who gave him a son named Ninurta who became the rightful heir. Enki finally had a son with Ninki, his wife. This was Marduk. Nannar, also known as Sin, was the firstborn son of Enlil by his wife Ninlil. He later became sovereign over Sumer's best known city-state of Ur and was responsible for Ur's great prosperity.

Enlil was in possession of the Tablets of Destiny that gave him power over Earth. A god named Zu stole the tablets which gave him the power that had been possessed by Enlil. Enlil's son Ninurta later retrieved the tablets, resulting in the banishment of Zu. Zu had been in possession of a flying machine from which he engaged Ninurta in battle. Scholars believe Zu may have been a contender for the Enlilship of Earth.

Inanna was one of the most colorful gods of Sumer. She was a daughter of Utu, the sun god, and granddaughter of Enlil. To the Greeks she was known as Aphrodite and to the Romans she was Venus. She was goddess of war and love to the Sumerians. Her domain was a land east of Sumer known as Aratta, and scholars believe her domain to be also in the Indus Valley. Her desire was to be goddess of Uruk where she occupied Anu's temple and shared his bed.

Inanna wanted the divine formula for civilizations that Enki possessed. She tricked Enki by getting him drunk and parted with 100 of the formulae. After Enki awakened, he realized what she had done, but it was too late. Inanna was on her way back to Uruk in her flying machine.

Inanna had an infamous reputation for sleeping with various rulers to advance her position and get what she wanted. Her ultimate goal was to become a member of the Pantheon of Twelve Gods, something she finally accomplished. Inanna was cunning, beautiful, ruthless, and often depicted as naked by the Sumerians. The Sumerian writings describe Inanna as flying from place to place in her "boat of heaven," which she often flew by herself, but normally her pilot/navigator was Nungal. The ruler of Uruk,was Enmerkar, who sent Inanna to Aratta to convince them to surrender to Uruk without bloodshed.

Inanna instituted the custom of the "sacred marriage" which was a sexual ritual where the priest or king was supposed to be her spouse for only one night. Enmerkar was the first of many invited by Inanna to share her bed under the guise of "sacred marriage." One of the most famous rulers of Erech, Gilgamesh, two-thirds god and one-third man, also had a love affair with Inanna. He became famous for his quest for immortality.

Sumer, around 2400 B.C., had become a country in need of strong leadership because of all the turmoil between the city-states. Inanna discovered Sargon I, a Semite, who was the perfect person to fill the role. He built a capital near Babylon called Agade. Sargon satisfied both Inanna's bedtime and political ambitions as she used Sargon to create an empire for her. A temple was erected for Inanna in Agade that served as her noble abode.

Historical records of Sargon's many conquests describe Inanna as being present on the battlefield. The empire spread well beyond Sumer, but he was not given Tilmun, the land of the gods. He also avoided Babylon, controlled by Ninurta but claimed by Marduk. Sargon ruled for 54 years and after his death, Inanna put one son after another on the throne.

For several years Inanna created havoc in her conflict with Marduk of Babylon, as there was no love lost between them. She had married a god named Dumuzi, a son of Enki, of whom Marduk disapproved. Inanna had descended to the netherworld to visit her sister and her husband was murdered in her absence. She blamed Marduk for the Dumuzi's death and she wanted revenge. Inanna persuaded the gods to bury Marduk alive inside the Great Pyramid without food and water. After a frightful time inside the pyramid, Marduk was eventually rescued.

Marduk returned from exile to Babylonia and fortified it to make it impervious to attack, enhancing the underground water system to help him withstand a prolonged siege. The Anunnaki gods did not want to use force to remove Marduk from the divine seat in Babylonia, so they sent Nergal, Marduk's brother, to convince Marduk to leave Babylonia. However, Nergal entered a forbidden underground chamber after Marduk's departure and removed a radioactive source of energy, a "brilliant weapon," also mentioned in the Old Testament. As a result, the day turned dark, flooding occurred, the land was laid to waste, and the gods became angry. This ended the conflict with Inanna and Marduk, and she made peace with Nergal.

Inanna enthroned her son Naram-Sin who commenced to conquer city after city, and he entered the Sinai where the spaceport was located. An assembly of gods was called to deal with Inanna's exploits. Fearing she was going to be seized, Inanna fled Agade for a period of seven years. She had deliberately defied the authority of Anu and Enlil by asking her son Naram-Sin to dismantle E-Anna (House of Anu) in Erech and to attack Enlil's temple in Nippur. The angry Enlil ordered hordes of Gutiums to lay waste to Agade, which they controlled for 91years. Inanna's mother and father took her back to Sumer and the era of Inanna was over.

After the nuclear catastrophe befell Sumer, it took several centuries for the land to recover and be resettled. The power had shifted northward to Babylon, and a new empire was to rise with the ambitious Marduk as the supreme deity. He had restored his position as the national god of Sumer and Akkad at the beginning of the second millennium B.C. The other gods were required to pledge allegiance to him and were to reside in Babylonia where their activities could be supervised. Marduk had usurped the Enlilship of the planet and initiated a large scale forgery of documents to make it appear that he, Marduk, was the Lord of Heaven, the Creator, and the Benefactor instead of Anu or Enlil. Marduk assigned himself as the deity of Nibiru, and the planet became known as Marduk to the Babylonians. The Babylonian *Epic of Creation* credited Marduk instead of Nibiru with crashing into Tiamet to create the planet Earth.

EGYPT AND THE SUMERIAN GODS

The gods of Sumer had great influence over Egypt's civilization. Tilmun, the land of the gods and location of the new spaceport, was located on the Sinai peninsula following the Deluge. According to Mantheo, the ancient Egyptian historian, the Egyptian god Ptah had reigned over the lands of the Nile 17,900 years before Menes of the First Dynasty or about 21,000 B.C. Nine thousand years later, Ptah handed over his rule to his son Ra. About 11,000 B.C., the rule of Ra was interrupted for about 1,000 years. This would have corresponded to the time of the Flood. Egyptians believed Ptah had returned to Egypt to help in its reclamation after the Deluge. Sitchin believes that Ptah was none other than Enki who had divided Africa between his six sons when the lands became habitable again.

Marduk was Enki's firstborn son and legal heir, whom some scholars believe was the Egyptian god Osiris. Sitchin believes Marduk was actually Ra, Egypt's sun god. Marduk rose to prominence when the Great Pyramids played an important role in his career.

Sitchin believes it was the Anunnaki who built the pyramids some 10,000 years ago in conjunction with the spaceport located on the Sinai. Egyptologists believe the pyramids were built during the Sixth Dynasty by Khufu (Cheops), by his successor Chefra (Chephren), and the third pyramid by Menkara (Mycinus). Sitchin provides proof that the Great Pyramids and the Sphinx already existed when kingship began in Egypt and that they were known in Mesopotamia. According to the Sumerian ancients, the Giza Pyramid and the Sphinx were landmarks for space-craft in the Sinai. The Great Pyramid contained instrumentation that helped guide the shuttle craft. An inscription on a stele near the Sphinx credits Ra as the engineer, who built the protected place in the sacred spaceport (Tilmun) from which he could ascend beautifully and traverse the skies.

Mesopotamian texts describe the Pyramid Wars where an agreement was finally made between Enlil and Enki to end hostilities and allow peace on Earth. The followers of Enki were no longer to inhabit the lands of Enlil. In exchange, the sovereignty of Enki and his descendants over the Giza complex had to be recognized for all time. Enlil agreed with one condition. Marduk, who had brought on the Pyramid Wars and used the Great Pyramid for combat purposes, would be barred from ruling Giza. Enki agreed to the proposal.

Enki appointed a young son of his to be Lord of Giza and Lower Egypt. His title was NIN.GISH.ZI.DA who became guardian of the secrets of the pyramids. This god was Thoth, who had married Enki's daughter, who was begotten after Enki made love to his sister Ninhursag. Manetho writes that it was Thoth who replaced Horus on the throne of Egypt about 8670 B.C. when the Second Pyramid War had ended. Thoth's reign was a peaceful time for Egypt, when the Anunnaki established settlements relative to the new space facilities. Mantheo assigned a reign of 3,650 years to the demigods who belonged to the Dynasty of Thoth. Thoth departed Egypt during the First Intermediate of Egypt (2160 to 2040 B.C.) when Ninurta asked Thoth to erect a temple for him at Lagesh. It was to be built from the humble dry clay of Mesopotamia, not of stone. This was at the time when the worship of Osiris and Horus was abandoned in Egypt, and the capital of Egypt was moved from Memphis to Heliopolis.

Enlil and Enki had come to another agreement. Enki and his sons would be allowed to come back freely to Sumer if Enki returned the site of Eridu to Enlil's control. Enlil agreed, and in return for his hospitality, Enki would bring prosperity to the land of Mesopoatamia.

From that day on, the land of the spaceport was known as Sin's land, the Sinai Peninsula. The command of the new Mission Control Center after the Flood was given to Sin's son Shamash and was to be located in Jerusalem (Ur-Shulim), Shulim meaning the "supreme place of the four regions." It replaced Nippur as the Mission Control Center and acquired the title as being "the naval of the Earth." A new beacon city, Annu, located outside of today's Cairo, was built and renamed Heliopolis by the Greeks.

Evidence has been presented that Sumerian legend depicted their civilization beginning with an influx of people from the sea. The Atlanteans, Sumerians, and Egyptians all claimed their prediluvian rulers as gods. The gods of Atlantis were similar to the gods of Greece, and scholars acknowledge that Greek mythology originated in Sumer. One must assume that during antediluvian times when Atlantis existed, the gods of Atlantis were the gods of Sumer. Egyptian legend describes their early rulers as gods who originated in both Atlantis and Mesopotamia.

The writings of Sumer date the creation of man approximately 300,000 years ago, while the ancient Nacaal writings assert the civilization of Mu (Lemuria) began more than 200,000 years ago. Archaeological finds in Africa date *homo sapiens* at approximately 250,000 years which is the earliest evidence of modern man. If the Nefilim arrived on the planet 432,000 years before the Deluge, Mu was a continent in the Pacific Ocean with evidence suggesting that it was the "Mother of all Civilizations." The Sumerian word Mu refers to the heavenly journeys of the gods or to the instances where mortals ascended to the heavens. Sitchin equates Mu with the Semitic word "shem" meaning "sky vehicle." Evidence links Mu, Atlantis, Sumer, and now Egypt.

THE ANUNNAKI ROOTS

One of the many mysteries about the Anunnaki stems from what star system did they originate. Some evidence suggests their roots were from Sirius, other evidence suggests Reptilians, and one source claims the Pleiades. The *Voyager* Series by Anna Hayes* claims they are from Sirius A.

Hayes' Ranthika extraterrestrial source claims that the Anunnaki were involved in one of the earliest seedings of Earth during the time of Atlantia, about 950,000 years ago. The Anunnaki brought to Earth the creed of the Templar Solar Initiates from Tara, which was "an elitist materialistic distortion of the principle of the Law of One . . ." The Anunnaki put a sexist view on the teachings, so women were viewed as being subservient to men, in order to use them as breeders for hybrid children. The teachings disempowered women so they would give in to sexual advances of the Anunnaki males. The

* Anna Hayes is now writing under the name of Ashayana Deane

Anunnaki created teachings that made them appear as gods, and they were able to manipulate the Atlanian cultures into submission. Hayes' source claims the Atlanian-Anunnaki hybrid were called the Nephilim, which quickly dominated the less-developed humans. Around 950,000 years ago, the Elohim and Ra confederacy removed the distorted Nephilim race from their connection to the Amenti morphogenetic field.

During the second seeding of Earth, the Anunnaki from Sirius A devised a plan to destroy the sphere of Amenti and to use Earth humans as a worker race to harvest Earth's gold for the Anunnaki purpose of replenishing the depleted gold in their universe. Also during the second seeding, the Elohim motivated Seres from a Harmonic Universe to interbreed with the members of the Atlanian and Aryan races to create a superior Guardian race who later became the Egyptian race. This hybrid race was known as the Serres and became the Pharaonic line of Egyptian culture. We know today the Egyptians carry the most diversified genetic code of all the races.

During the second seeding era, a 1,000 year war broke out between the Elohim and Anunnaki for control of the Earth and for the direction of human evolution. Most races on Earth sought exile in the other planetary systems, while others retreated to the Inner Earth. Most of the Nephilim relocated to Sirus A and then to the planet Nibiru. Around 848,000 years ago, the treaty of El Annu was reached when the Nephilim agreed not to return to Earth.

A group of Anunnaki did not agree with the treaty and they joined forces with the remaining Draco hybrids to try to destroy the Sphere of Amenti. They were called the Anunnaki Resistance. The Sirian Council along with other civilizations enacted the "The Templar-Axion Seal" (666) to the Anunnaki Resistance.

During the third seeding of Earth, the Anunnaki had plans to destroy the Sphere of Amenti in order to use humans as a working race. The Guardians removed the Sphere of Amenti and placed it in the Andromeda galaxy, and a Sirian-Arcturian coalition for Interplanetary Defense was formed to protect the Sphere of Amenti. The Anunnaki Resistance and Draco Allies began to infiltrate the Atlantean culture about 55,000 years ago, and the Law of One teachings were once again distorted. The Anunnaki-Annu gained control over Egyptian land. It was the action of the Anunnaki Resistance and Dracos that resulted in tremendous explosions from the overload of generator crystals causing the destruction of Lemuria and later the Atlantis civilization.

During that time period, a chasm developed between the races that followed the Law of One and those who were influenced by the Anunnaki Resistance Templar Solar Initiates creed. The Templar-Annu began to dominate the Atlantean culture and made plans to conquer Egypt.

Pyramids were a hallmark of Anunnaki architecture. Some of the small pyramids were built over the portal passage area through Egypt. The Great Pyramid was built over the main Egyptian opening to the Inner Earth through which the Ark of the Covenant could be accessed. The original Sphinx covered the portal to the Inner Earth and was connected to the portal passage leading to the Ark of the Covenant.

The crystal power generator was exploded by the Templar-Annu, which ripped apart the Atlantean continent, causing it to sink in 28,000 B.C. Before the explosion, the Sirian Council removed various groups to Egypt and Inner Earth.

The Anunnaki Resistance launched an attack against the Sirian Council outpost on Mars and tried to destroy the Ark of the Covenant in Egypt. The Sirian Council drove the Anunnaki Resistance out of Earth and back to planet Nibiru after the destruction of the Martian outpost.

Mark Pinkham, in his book *The Return of the Serpents of Wisdom*, tells of Sumerian texts that "mention the periodic visitations of wise serpents called Anunnaki, some of whom may have come from Sirius." Robert Temple in his book *The Sirian Mystery* also believes the Anunnaki came from Sirius. Temple writes that the Sumerian Anunnaki are referred to as the "sons of Anu" and believes that Anu may be a star in Sirius.

Enki's symbol was the goat fish who arrived in Mesopotamia with his Anunnaki entourage. Greek translation says he was called "repulsive" or "dragon faced." In Sumerian texts, the

Kings are collectively referred to as a lineage of ten preflood priests known as AB-GAL or "masters of knowledge." They are depicted with fish-like bodies. Enki was also depicted with a fishtail.

Some researchers conclude that a wave of Anunnaki came from Sirius while some Sumerian texts suggest that some delegations may have arrived from Atlantis. Evidence also suggests they had fish-like qualities of the Sirians, with some clues suggesting they were Reptilians. Could they have been Reptilians from Sirius, not Reptilians from Orion?

Chapter Eight

OTHER GALACTIC CIVILIZATIONS
From Blonds to Praying Mantis

A number of galactic civilizations interact with humans at some level, but not all of them have tried to communicate directly with humans. There are reports by some abductees who remember various extraterrestrial races being present during their abduction. Those individuals who have communicated with the Pleiadians, Sirians, and Zetas have been told about other ET races who are trying to help humanity on its evolutionary path. This chapter will try to piece together bits of information that discuss these various galactic races.

The Blondes

One of the most common races seen during the abduction experience are the Blondes, sometimes called the Nordics or Swedes. Human-looking Blondes have been seen in about 60 percent of the abductions. George Andrews discusses this race in his book *Extraterrestrials, Friends and Foes*. The Blondes are humanoid, stand between 6 and 6-1/2 feet in height and have blonde hair and blue eyes. They come from a planet revolving around Procyon, a binary yellowish-white star system that rises before Sirius in Canis Minoris, located about 11.4 light years from Earth. Blondes abide by the galactic law of noninterference and will not help Earth, except in a few special situations. They will interfere only if a Gray activity adversely affects another part of the universe. Blondes refuse to give governments advanced military technology, but they do want to warn us about the negative Grays. They have an underground base located near Carlsbad Caverns in New Mexico.

159

For thousands of years, the Blondes have been arch enemies with the negative Grays from the star system Rigel in Orion. Their attitude appears to be benign toward humans except in cases of inhumanity toward one another. They are also upset with our government for having made a secret alliance with the negative Grays in order to obtain military technology. Because the Procyonians will not provide military technology, our government is not interested in dealing with them.

Their greatest concern is the relationship the negative Gray has with our government. They encourage us to regain control of our government to extricate ourselves from the covert alliance between the CIA and the Grays. They also caution that one can never defeat or control anyone except oneself. To overcome the Rigelians (negative Grays), one must take great precaution not to become the enemy ourselves. They say war is neither right nor wrong, as wisdom can be gained on any path, similar to the Law of One philosophy.

Much of humanity's evolution can be attributed to the Blondes. Their DNA was combined with the Neanderthal, resulting in the sudden emergence of the Cro-Magnon human. The genetic engineering combined the larger brain capacity of the tall Blondes with the lung capacity and respiratory systems of the primitive human. Procyonians have continued to crossbreed at many stages in our evolution to help our evolutionary development, and it still goes on. Their motivation for breeding with humans is to assist humans in helping themselves.

The Procyonians have the ability to tune in on humans telepathically to vicariously experience our emotions. Their purpose is to study and understand the biological and psychological ramifications of emotions. Over all, they are concerned about the well-being for all life forms, not just humans, and with what we are doing to our environment.

The Blondes have evolved to where they can travel through space and time. Through thoughts, they can teleport themselves. On the physical level, they do have physical space-craft.

Historically, the ancestors of the short Grays were once the tall Blondes. Following the nuclear Great War, when the Rigelians had become the short Grays, it took thousands of years to reconstruct their society. The Procyons had been spared the chromosome and glandular damage experienced by those who stayed behind on Rigel, as all were transformed into short Grays. The Blondes claim there are still two wars going on. One war is between Rigel and Procyon, which is in a state of temporary truce. The other war is between Rigel and Sirius that is being fought actively. The Blondes assert that Earth was seeded by the tall Blonde Rigelians prior to the Great War. Because of the common ancestry, the Blondes and Grays have a great interest in Earth humans.

Both the Blondes and Grays have the ability to disintegrate matter into energy and reintegrate it back into matter. This allows an abductee to be transported through walls and roofs. The Grays also have the ability to camouflage themselves as tall Blondes through mental energy projection. In earlier centuries, Blondes were sometimes mistaken for angels. They seem not to age and appear to be from 27 to 35 years in human age, no matter their real age.

Blondes are considered the ideal extraterrestrial to interbreed with humans of all races. Humans that interbreed with other extraterrestrials must be carefully monitored, because humans have a different immune system from extraterrestrials. Any new hybrid bacterial strain can cause disaster in both races. Blondes do have the ability to infect humans. The Blondes claim that there are over 1,000 humans in the United States who are offspring of galactic beings and Earth humans. They also emphasize that all humans have extraterrestrial genetic background. Through recorded history, the Blondes have constantly manipulated our genes to breed out negative traits.

Eons ago, according to the Blondes, the Grays wanted to make peace with Procyon and normalize relations. They began to visit Procyon as tourists and then infiltrated society at all levels. Through telepathic hypnosis, the Grays were able to control directors of their intelligence systems, the leaders of their planet, and most of the Blondes. It was like a spell was cast on Procyon, as if they were being programmed by black magic. Some Blondes left Procyon just before the great War broke out, but those who stayed were under the dominion of the Grays.

The Blondes who are working on board the space-craft with the Grays are hybrids or clones. The authentic Blondes have distinct feature differences and do not look alike. They are muscular with slender necks and agile bodies. The clones have thick necks and coarse muscular bodies and are given orders by telepathy. Authentic Blondes are almost identical to earth humans and actually have stronger immune systems than humans. Because of their past with the Rigelian Grays, no true Blonde would collaborate voluntarily, but some Blondes have been taken prisoner of war and have no choice. Humans are being warned by the Blondes about what happened to their culture because it is now happening to ours.

For Earth humans to fight Grays on a military level is nearly impossible. Their technology is advanced far beyond ours. For example, they have electromagnetic technology that can switch Earth's axis tilt. The Blondes say the only way to become victorious over the Grays is to change our consciousness. For those who are being abducted, they should try not to be afraid but put their minds elsewhere by focusing attention on dynamic protection imagery of a religious or mystical nature. No matter what the physical situation is, it is essential to send telepathic messages for protection from higher powers without fear, anger, or hatred. If we act toward the Grays out of hatred, we become like them, as negative emotions bind us to them, and the Grays anticipate and encourage the negative emotions so that they can manipulate us. Grays are acting in desperation because their race is terminally ill with exhausted DNA, and their bodies degenerate from generation to generation. The only reason the Grays can dominate behind the scenes is that the Secret Government on Earth and a few elected officials have made deals with them.

According to the Blondes, since antiquity our planet is divided into four sections between the Blondes, Grays, Arcturians, and Reptilians. These groups still consider themselves owners of the planet. However, an alliance has been made between the Blondes and Reptilians to drive out the Grays. The Blondes claim that the Grays are partially Reptilian themselves and know how to manipulate the reptilian level of Earth humans' brains. They can temporarily paralyze all portions of the mammalian brain higher than the "ape level." When the "ape level" of our brain is activated, our brain creates violence, greed, lust, and rage. Any attempt to fight back at the "ape level" ensures the success of the Grays. Abiding by the galactic noninterference policy, the Blondes are here to help Earth humans. They know what happened to their planet because of the negative Grays, and they want to prevent it from happening to Earth.

The Ranthians

The Ranthun Keana race is highly evolved both from a technological and spiritual standpoint, and they are referred to as Ranthias. They come from a distant star system of our universe and have mastery over the time and space portals. Most of the information received by Anna Hayes* in the *Voyager* series came from the Ranthians. They told Anna they have been involved with Earth since its beginning and have been our guardians in numerous ways, protecting Earth humans and our planet from interference

Ranthun Keana

by multidimensional beings and galactic beings. Their mission is to assist and offer knowledge and possibly intervene in the Zeta's plan, if possible.

The Ranthians can park between dimensions and appear invisible to humans. Unlike the Zetas (Grays), the Ranthians are not completely based in matter. They have the ability to manifest in physical forms in matter densities similar to humans. Ranthian beings are masters of the hologram, simialr to the Zetas. Their "organic form" has properties of fluidity associated with water. Because of this fluid form, they can take on many shapes and are good at shape-shifting. Ranthian beings

* Anna Hayes is now writing under the name of Ashayana Deane

are tall and thin, emitting light through translucent white skin. They have almond-shaped eyes of various colors with kinky white hair. Spiritually, they are very advanced.

The Ranthians are very skilled at making their presence known through light, as they create manifestations in conjunction with light phenomena. They are masters of illusion, but don't create illusions of Three-Dimensional solid objects. Ranthians understand the mechanics of the human bioelectric circuitry and the electromagnetic properties of earth. They mainly interact with humans during the human dream state and altered consciousness states.

Ranthians are aware of the Zeta agenda and are attempting to teach humans, Zetas, and other races about ideologies and technologies that will help them collectively work toward a peaceful evolution for all concerned. For those who desire to learn, they provide information and assistance. However, they must be asked to intervene, and they will not impose their assistance or contact upon humans without invitation. Their ethical code honors free will of all life.

Ranthians are considered the Guardians and overseers of the Time Portal Project. They are closely related to "keepers of the time" Maya in our time system. They and other portal guardian groups are beginning to interface with our time system again. Both the Ranthian and Aethian races are working together to educate humans for what is necessary on humanity's part to fulfill their evolutionary path on Earth.

Aethiens

Many abductees have reported seeing strange beings that resemble upright preying mantises on board space-craft. These are the Aethiens who are large, white, graceful beings that are developed spiritually. The Aethiens always work as teachers. Often they accompany positive Zetas, who have agreed to work in brotherhood with all of life by assisting humans in their encounters with the Zetas. Both the Aethiens and Zetas have the ability to shape-shift when dealing with humans and can take on an animal or human form, according to the *Voyager Series* by Anna Hayes*.

The Aethiens originate from a galaxy that exists within a dimension adjacent to earth, and they fall into a category of being interdimensional extraterrestrials. By using the time portal system, they can work with humans in our dimension. They come as emissaries of peace and growth toward brotherhood for all species.

The Aethiens and several other interdimensional races agreed to assist the Zetas in creating a mutation in their lineage through genetic engineering. They also worked with the Zetas in creating a being called the Zionite, using a combination of Aethien, Zeta, and human DNA. There are several planets in our universe where Zionites can be found, and several large groups of Zionites are working with the Aethiens on their home planet. Zionites exist as a part of our future in our Three Dimensional world, and they are part of human heritage.

As will be discussed in a later chapter, humans originated on Earth's Fifth Dimensional counterpart planet Tara. Zionites were one of the star groups whose genetic material was supplied to create the original human prototype, a race called Turaneusaim, which emerged 560 million years ago. The Zionites interacted with the Sumerian and ancient Egyptian cultures. The gentic package found in Zionites is called the silicate matrix or crystal gene. The silicate matrix contains the original twelve-stranded DNA code of the original Taran-human prototype, the Turaneusiam. The DNA exists within a number of humans as a latent genetic code sequence that must be activated. Once activated, it will allow for progressive transformation of form. Humans who carry the matrix have the potential for accelerated evolution. Those with the genetic makeup will one day be able to traverse the time portals and interdimensional passageways without deterioration of biological form. They then will have predisposition to multidimensional perception and interdimensional communication. There is a time in Earth's future that we will have direct contact with the Zionites. These future events are the goals lying behind the genetic experiments carried out by the Zetas, Aethiens, and humans with soul agreements.

A group of extraterrestrials called the Watchers have formed an alliance with other ET civilizations to help as many humans as possible who request assistance. During the mid-1980s, they had agreed that humans had evolved enough to warrant outside intervention. The Watchers enlisted the help of the Aethiens, Ranthians, Ashtar, and other groups to accelerate spiritual evolution so humans would not resist with malice as they encountered a Zeta.

Breneau

This race of advanced beings belong to the Guardian Alliance of benevolent extraterrestrials that are here to assist Earth's human evolution. From the highest dimension, they appear as tall luminescent figures that have elongated heads and large eyes when they physically manifest.

Andromedans

Alex Collier has been in frequent contact with the Andromedans and has written several books about his experience and knowledge he has received from them. The name of his Andromedan contact is Moraney. The Andromedans are a highly evolved civilization and are here to help the evolution of humans. *Wikepedia* describes the Andromedans as "arch angels" and "omnipresent light beings" who help oversee human evolution.

The Andromedans are waiting for 10 percent of Earth's population to awaken consciously and to be asked for some kind of intervention. They think of themselves as healers and carry a strong pride in their technology in the arts of healing the physical, emotional and spiritual bodies. They show others how to integrate their belief systems and feelings to resolve conflicts. In the ancient past they have helped Earth resolve serious conflicts, and their efforts have done much to raise the overall consciousness in our universe.

They formed the Andromedan Council which consists of 140 various star systems whose purpose is to deliberate upon Earth. Because Earth has been manipulated for over 5,700 years, they feel humans deserve a chance to prove themselves. Their chief activity is to facilitate decisions of the Galactic community. They promote the growth of psychic/crystal/rainbow children, peace, education, exposing elite manipulation, and promoting global governance, diplomacy, and conflict resolution. They also promote social justice, human freedom, and responsible use of advance technology.

The strongest message they want humans to know about is that humans have to do their own spiritual work and not depend on a Messiah. Humans have to be their own savior as they are only responsible to themselves.

Andromedans come from a star system they call Zenatae in the Andromea Galaxy. Many planets in the Andromeda system have water with some planets covered entirely by water. Some Andromedans live on the surface of the water and some below. In Andromeda, Alex was told, the Andromedans teach their children everything, and the oldest and wisest Andromedans are the teachers, that hold nothing back.

Alex Collier was told by Moraney that the negative Grays gave HIV virus technology to NSA, which handed it down to the military complex. Africa was used as a testing ground for AIDS. The Andromedans claim there are 15,000 Grays and 1,833 Reptilians living in bases beneath the United States, and they are using 72,000 walk-in souls to help awaken humanity. The Andromedans confirm Preston Nichol's claim that humans from Sirius B provided technology for time/interdimensional in both the Philadelphia Experiment and Montauk Project that will be discussed later. A whistle blower has confirmed this saying the NSA has a clandestine organization ongoing today called the Advanced Contact Intelligence Organization regarding time travel. Sirius B also gave technology to NSA that can really mess with our minds. According to the Andromedans, Sirius A visitors were the original builders of our grid, the architecture on which our planet is built based on sacred geometry.

Andromedans are closely associated with the races of Tau Ceti and Cygnus Alpha, with the latter race knowing a great amount about acoustic technology and sound. Collier was told that the Alpha Centauian race has played a significant role in human affairs by promoting social justice, human freedom, and responsible use of advanced technology.

The Andromedans gave Alex hope against the adversarial extraterrestrials and said by creating a space of love, the Grays, Dracons, and negative Orions won't be able to tolerate the higher vibration.

Lyrans

In 1977, Billy Meier was contacted by a galactic visitor from Lyra who had a Nordic apppearance. Meier's Swiss Pleiadian contact told him that both the Pleiades' and Earth's ancient ancestors originated near the Ring Nebula of Lyra, again confirming other sources that Lyra was the original seed of humanity.

According the Michael Salla, Ph.D. in his book *Exopolitics,* Lyrans are considered to be the Galactic historians for the human species. Their primary activity is to record and analyze the unique history of the human race in the galaxy and to assist in the understanding of human motivation and potential.

Lyrans are the original white Aryan race, Nordic in appearance. The Pleiadian and Andromedan races have preserved the true Aryan lineage.

Lyra is a dim constellation which is home to the brightest star in the sky, Vega, and consists of 14 inhabited planets. Some other races that evolved from Lyra include Cignus Alphan, Alpha Centauri, Cassiopia, and Sagittarius A and B.

Vegans

Not to be out done by the Lyrans, the Vegans also contacted Billy Meier. They originate from within the constellation of Lyra. Vegans are a darker or blue skinned race who are considered by some to be the blue race of advanced humans mentioned in the Vedic texts and from whom the Hindu gods Vishnu and Krishna were derived. Jefferson Souza, a Brazilian contactee, met the Vegans of Lyra and described them to be similar to the natives of India.

Alex Collier was told that Vega was the first star system in Lyra to be populated by humans. They have played an important role in colonizing the Galaxy. The first Lyrians, together with the Vegans, came to Earth initially 22 million years ago.

Tau Cetians

Another humanoid race that visits Earth comes from the star system Tau Ceti and Epsilon Ericlani. This race is white in color, whose main purpose is to help Earth humans to deal with the covert strategies and activities of the negative Grays.

One of their space-craft crashed resulting in the capture of a Tau Cetian astronaut by the U.S. military. According to a whistle blower who claimed the Tau Cetian was tortured by its captors, it nearly caused an incident with other Tau Cetians who wanted to intervene militarily.

Because the Tau Cetians had been victimized by the negative Grays, they are now working with other races and communities who are also victims. Their objective is to raise awareness about negative extraterrestrials, to identify corrupt elites and institutions, to develop strategies in dealing with this diabolical force and their advanced mind control technologies. Essentially they are here to raise human consciousness.

Ummites

The Spanish were blessed by a group of visiting extraterrestrials from a planet Ummo, located 14.6 light years away, thought to be in the star system Wolf 424. The Ummites had a secret hidden base near a small town in the French Alps. During the 1960s and 1970s, the Ummites played a significant role in disseminating scientific and technical literature in Spain, which was shared with the rest of Europe. This scientific knowledge and theory was used to address global problems.

They made contact through phone calls and regular mail. Their front was an Ummite, who portrayed himself as a Danish doctor who hired a typist to disseminate the information. The typist eventually discovered the true identity of her employer.

On June 10, 1967, an Ummo space-craft landed near Madrid after announcing the time of their landing to a select group of UFO researchers. Because of a potential nuclear war in 1973, the Ummites terminated their residence on Earth.

Queventelliue

Another Guardian race that helps humanity is called the Queventelliue. On Earth we call them Sasquatch or Bigfoot or Yeti. They are large, tall (around seven feet), long-haired, apelike beings of great intelligence and sensitivity. They are occasionally seen on Earth as they monitor Earth's environment for guardian purposes. Those who have encountered them at close range report a horrific odor. One Sasquatch researcher, Jack Lapseritus, has lived among a family of them in Oregon. He confirms that they are interdimensional and that UFO sightings are common in the area of Sasquatch sightings.

The Borentulism Race

This race is a pure strain of a human-Zeta hybrid, referred to as hybrid whites. They are created through fetal transplants and have the ability to enter into Three-Dimensional frequency for a short period of time. The Borentulism hybrid has a much stronger human imprint as compared to an egg donor conception called a Borentasai hybrid. Some day in the future they will serve as an intermediary between humans and Zionites, according to Anna Hayes'* Ranthian source.

The white hybrids created from fetal transplants are capable of limited interdimensional and transmigration. The egg-donor white hybrid cannot endure interdimensional transplant. Both strains have the ability to link their minds directly with human contactees using the Keylonta codes. They also interact with humans during the dream state.

Beta Centauri (Hyperboreans)

According to Lyssa Royal's extraterrestrial source, there was a group of scientists on Earth eons ago who tried to control others. The people on Earth exiled these scientists to Beta Centauri, which was the closest sun system to Earth located 4.3 light years away. The group of scientists, led by a person named Arus , plotted to return to Earth and attack it in order to seek revenge for their exile. On their return to Earth, they took over a small region of land they called Hyperborea, which is today's Florida.

In 13,000 B.C., a person named Samjase, second in command to Arus, performed genetic mixing experiments with animals and less developed Earth humans in order to try to improve man. Arus and his followers started small wars on Earth, and his son Arus II took control of Arya or what is now India, Pakistan, and Persia.

According to Royal's source, about 9498 B.C., during the end times of Atlantis caused by an exploding asteroid, the Earth rolled on its axis. Florida had been located where Greenland is today. Hyperborean survivors had to move underground and now live beneath Mount Shasta. There have been many current day anecdotes about people encountering humans that live within the earth under Mount Shasta. Books have been written about these underground people, but many sources claim these underground inhabitants were the legacy from Lemuria. However,

Royal's source claims they are the descendants of the Hyperboreans who came to Earth 30,000 years ago. The main entrance to this underground civilization is located on the east side of Mount Shasta. They possess an advanced technology and their space-craft are gold in color, which accounts for the many UFO sightings are reported in the Mount Shasta region.

The Ashtar Command

Many New Agers interested in ufology are familiar with the Ashstar Command, whose message is similar to those of other galactic civilizations we have discussed, with an emphasis on spiritual principles and our evolvement to the Fifth Dimension.

Contact with the Ashtar Command began on July 18, 1952 when they contacted aviator Col. Van Tassel who wrote six books about his experience, including one entitled *I Rode a Flying Saucer.* Tassel was told that the purpose of the Ashtar Command was to save mankind from itelf, warning of the dire consequences of nuclear weapons.

Tassel's purpose was to build receptivity to the Ashtar communication. He was told that mankind was tinkering with a formula it does not understand, referring to exploding the hydrogen atom. Members of the Ashtar command would be monitoring Earth to prevent nuclear wars.

Over a half-century later an article appeared in the *India Daily* on February 20, 2005 saying that Indian scientists are slowly understanding that extraterrestrials have a very unique power of jamming the operational systems of any manmade device. They can easily jam the operations of any nuclear missile. In recent days, both India and Pakistan found their nuclear missiles disabled, as had the Americans and Russians in previous years. The Indian scientists said that the extraterrestrials will disable nuclear weapons if there is any immediate threat of war.

Who is the Ashstar Command? They are a Brotherhood of Light under the spiritual leadership of Prince Sanada, the galactic name of Jesus the Christ. They are here to assist the planet Earth and humanity in this cycle of cleansing and realignment, writes author Jose Garcia. Their message is spirituality that focuses on our Divine Nature. They are also known as the Galactic Command and Command of the Extraterrestrial Force of the Intergalactic Federation.

Their commander is named Sheran who is leader of the Cosmic Plan for our transition into the Fifth Dimension. His command ship is named the Eagle. Under his command are 10 million highly evolved spiritual beings that surround our galaxy in a protective electronic circle within the Alliance for Peace in the Intergalactic Council. Lord Ashstar is now an immortalized soul, spiritual traveler, and Ascended Master, and works through Archangels Michael, Gabriel, and Sanada/Jesus.

There are several million volunteers who have incarnated at this time to help prepare the planet for the transition to the Fifth Dimension. These star volunteers have incorporated in them genes that can be activated once they acquire a certain frequency to help with the Divine Plan. Their work is in coordination with the legions of the Archangels Michael, Uriel, Gabriel, and Jophiel and the 70 Brotherhoods of Light that administer the Divine Plan.

The Ashtar Command has different fleets that specialize in spiritual education, ascension, scientific investigation, communications, and the welfare of humanity. They also have the ability to deflect incoming asteroids and the ability to diminish the tilting of the Earth's axis.

The Ashstar Command is the universal ambassador of peace who advocate unity, harmony, and the peaceful coexistence of all. Not included in the Command are extraterrestrials who are involved in abductions, implants, intimidation, fear, and menace. The Command adheres to principles of noninterference as established by the Galactic Federation respecting the right of an individual's free will.

Some of the major tasks of the Ashstar Command include: (1) Assisting with the energy level to shift Earth from the Third Dimension to the Fifth Dimension. (2) Maintaining the stability of the Earth's polar axis. (3) Monitoring the Earth's grid systems (4) Inspiring the spiritual expansion of consciousness. (5) Encouraging the shift from fossil fuels to free energy. (6) Safeguarding the service missions of the star volunteers. (7) Evacuating the world's population from a potential geophysical catastrophe. (8) Planning for the return of the Christ.

There are many other galactic civilizations that are interacting with earth and humans that we know very little about. Most of these civilizations are trying to help our spiritual evolement as they are undergoing evolution themselves.

Chapter Nine

THE INNER EARTH
CIVILIZATIONS
It's Heaven, Not Hell

Surprisingly, there are civilizations who live within the Earth, that are far more advanced than civilizations who live on its surface. How can this be? According to the beings who inhabit the Inner Earth, the Earth is hollow as are all other celestial planets, including the sun. There are openings to the Inner Earth located at both poles that are camouflaged by clouds and perhaps a hologram. Admiral Richard Byrd flew into the Inner Earth while exploring the area beyond the North Pole in 1947, where he encountered beings that gave him a message to take back to surface humans. Several areas of the Inner Earth are inhabited, including the subterranean Earth located in the crust not too far below the surface and the hollow Earth cavity in the middle of the planet. One of the civilizations living below Mount Shasta is that of the descendants from the survivors of Lemuria 12,000 years ago. They call their home Telos. The civilization occupying the interior hollow are called Catharians, whose seat of government is called Shamballa. Collectively, the civilizations living within the Earth are called Agartha. Only a few surface humans have been allowed to enter this amazing civilization, including Buddhist lamas from Tibet. To give credibility to this extraordinary claim, we need to examine the remarkable flight of Rear Admiral Byrd beyond the North Pole.

THE FLIGHT OF ADMIRAL BYRD

Admiral Byrd is one of America's best known explorers, exploring both the Arctic and Antarctic. In 1947, Byrd wanted to explore by air the region beyond the North Pole. Several books have been written about this journey including the book by Raymond Bernard, Ph.D. entitled *Hollow Earth*. During the February 19, 1947 flight, Byrd planned to fly 1,700 miles beyond the North Pole. As he progressed beyond the North Pole, he encountered iceless land and lakes, mountains covered with trees, and even a monstrous animal resembling a mammoth of antiquity moving through the underbrush, as reported by the plane's occupants. For nearly 1,700 miles the plane flew over land, mountains, trees, lakes, and rivers that should not have been there. In 1956, Byrd flew 2,300 miles beyond the South Pole about which he wrote, "Once again we have penetrated an unknown and mysterious land which does not appear on today's maps." Any further information about the flights was suppressed by the government. Years following Admiral Byrd's death, his family released Byrd's secret diary about his flight over the North Pole. Byrd said he must write this diary regarding his Arctic flight in secrecy and obscurity. He said he was "not at liberty to disclose the following documentation at this writing . . . perhaps it shall never see the light of public scrutiny, but I must do my duty and record for all to read one day. In a world of greed and exploitation of certain mankind, one can no longer suppress that which is truth."

His diary began at 0600 hours of February 19, 1947 following take-off. At 0910 hours his compass begins to gyrate and wobble, and at 0915 hours he sees mountains. Byrd reports seeing a green valley at 1000 hours and writes, "There should be no green valley below." At 1005 hours he sees a mammoth-like animal, and at 1030 hours he flies over more green hills and the external temperature reads 74 degrees Fahrenheit. Byrd spots a city at 1130 hours and reports, "off our port and starboard wings are strange types of aircraft. They are closing alongside."

The radio begins to crackle at 1135 hours and a voice comes through in English. The message is "Welcome, Admiral, to our domain. We shall land you in exactly seven minutes. Relax Admiral, you are in good hands." He notes that the engines have stopped running and the aircraft is under some strange control. The controls are useless. They touch down at 1140 hours with only a slight jolt. Several tall men with blonde hair approach the aircraft, and Byrd sees no weapons and is called by name to open the cargo door.

The radioman and Byrd were taken from the aircraft and received in a most cordial manner. They boarded a small platform conveyance with no wheels and were taken to a glowing city of crystal material. Both were given beverages in a very contemporary large building. The two aviators were then taken to another room and one of the hosts said, "Have no fear, Admiral, you are to have an audience with the master." He is greeted with a warm rich voice by the master, a man of delicate features and etching of years upon his face.

He said, "We have let you enter here because you are of noble character and well-known on the surface world. You are in the domain of Arianni, the Inner World of the Earth. We shall not long delay your mission. . . But now, Admiral, I shall tell you why you have been summoned here. Our interest rightly began just after your race exploded the first atomic bomb. . . We have never interfered before in your race's wars, and barbarity, but now we must, for you have learned to tamper with a certain power that is not for man, namely that of atomic energy. Our emissaries have already delivered messages to your world, and yet they do not heed. Now you have been chosen to be witnesses here that our world does exist. You see, our culture and science is many thousands of years beyond your race. . . . the dark ages that will come now for your race will cover the earth like a pall, but I believe that some of your race will live through the storm. We see at a great distance a new world stirring from the ruins of your race. When that time comes, we shall come forward again to revive your culture and your race . . . You, my son, are to return to the surface world with this message . . ." Following the meeting, Byrd and his radioman were ushered back to their plane by an escort who got them back on their way. They landed safely back at their base at 0300 hours.

On March 11, 1947 Byrd attended a staff meeting at the Pentagon but was further detained for 6 hours and 39 minutes when he fully stated his discovery and the message from the master. He was ordered to remain silent in regard to all that he had learned! Byrd was reminded that he was a military man and must obey orders. On December 30, 1956 Admiral Byrd made his final diary entry. He said he "faithfully kept the matter secret, which has been completely against my values and moral rights."

Confirming Admiral Byrd's discovery were several individuals who claimed that they had entered the North Polar opening. Several penetrated far enough to reach the subterranean world of the hollow interior. Dr. Nephi Cotton of Los Angeles reported about a patient of his who had lived in Norway near the Arctic Circle. He and a friend took a boat trip to go as far as they could into the North Country. At the end of one month, they sailed into a strange new world, a vast canyon leading into the interior of the Earth. They encountered warm weather and saw a sun shining inside the Earth.

They sailed up a river and saw the Interior Earth, which contained both land and water. They saw huge trees and then encountered giants dwelling in homes and towns. They were invited by the giants into their homes. For one year, the two sailors stayed with the giants and enjoyed their companionship, as the giants were never unfriendly to them.

Another anecdote involved Olaf Jansen and his father who were from Norway, and whose boat was blown off course into the Interior Earth. They spent two years there and returned through the South Polar opening. A book entitled *The Smokey God* was published in 1908 telling of their experience. They said the people lived from between 400 to 800 years and were highly advanced in science. The people could transmit their thoughts from one to another by a certain type of radiation. They reported a power source that was greater than their electricity on the Earth's surface. Another astounding claim was that the inhabitants were the creators of the flying saucers, which were operated by their superior power source, drawn from the electromagnetism of the atmosphere. The people were very tall, 12 feet and taller. All these anecdotal descriptions of the Interior Earth were similar, and all were independent of each other.

AGARTHA

Agartha is the name of the subterranean world within the Earth. The word Agartha is of Buddhist origin, and all true Buddhists believe in its existence. They also believe that this subterranean world, whose capital is Shamballah, has millions of inhabitants with many cities. Here dwells the "Supreme Ruler" of this empire known as "King of the World" in the Orient. According to many Buddhists, it is believed that he gave orders to the Dali Lama (before the Chinese occupation) who was the Inner Earth's terrestrial representative. His messages were transmitted by messengers through certain tunnels connecting the subterranean world with Tibet. According to the Buddhists, Agartha is a subterranean paradise, and it is the goal of all true Buddhists to reach it.

Nicholas Roerich, the famous Russian philosopher and explorer, claimed that Lhasa, the capital of Tibet, was connected to Shamballah with a tunnel. The entrance of the tunnel was guarded by lamas who kept its whereabouts secret as directed by the Dali Lama.

The Agartha civilization is believed to be descended from the Atlantean civilization which fled to the Inner Earth after learning about the futility of war, and they have lived in peace ever since. Buddhist legends state that Agartha was first colonized many thousands of years ago when a holy man led his followers underground.

Raymond Bernard, Ph.D. writes in *Hollow Earth* about a conversation a lama had with a Russian explorer named Ferdinand Ossendowski who wrote about the event in his book entitled *Beast, Men, and Gods*. The lama had been to Agartha and said, "All inhabitants of this region are protected against evil, and no crime exists within its boundaries. Science developed tranquility, uninterrupted by war and free from the spirit of destruction. Consequently, the subterranean people were able to achieve a much higher degree of wisdom. They compose a vast empire governed by the King of the World. He masters all the forces of nature, can read what is within the souls of all and in the great book of destiny. Invisibly, he rules over 800 million human beings, all willing to execute his orders.

"All the subterranean passages in the entire world lead to the world of Agartha. The lama said that all the subterranean cavities in America are inhabited by these people. The inhabitants of submerged prehistoric continents (Lemuria and Atlantis) found refuge and continued to live in the subterranean world.

"The capital of Agartha (Shamballah) is surrounded by villas where live the Holy Sages. . . . Their palace is surrounded by the palace of the Gurus, who control the visible and invisible forces of the earth from its interior to the sky, and are lords of life and death. If our crazy humanity will continue its wars, they may come to the surface and transform it into a desert . . . In strange vehicles, unknown above, they travel at unbelievable speeds through tunnels inside the earth.

"The lama said that a great number of people have visited Agartha but most who were there maintain the secret as long as they live.

"Many times did the rulers of Ourga and Lhasa send ambassadors to the King of the World," said the lama librarian, "but they could not reach him. However a Tibetan chief, after a battle with the Olets, came to a cavern whose opening bore the following inscription: 'This door leads to Agartha.'

"From the caverns left a man of beautiful appearance, who presented him a Golden Tablet bearing a strange inscription saying, 'The King of the World will appear to all men when comes the time of the war of the good against evil, but the time has not come yet. The worst members of the human race have yet to be born.'"

South America

Besides Tibet, Brazil has similar mysterious tunnels that honeycomb the area below its surface. These tunnels serve as a contact with the subterranean world of Agartha. The Brazilian tunnels are thought to have been dug by the Atlanteans, who had colonized Brazil and found refuge underground when Atlantis was undergoing destruction millennia ago.

The most famous Brazilian tunnel is located in the Roncador Mountains of northwest Matto Grosso. Fierce Chavantes Indians guard the opening of the Roncador tunnel. Legends say they will kill anyone who dares to enter the tunnel uninvited and who might molest the subterranean dwellers whom they respect and revere. The Murcego Indians are another tribe that guard secret tunnel openings. Archeologists are puzzled by the enigma of these tunnels.

A number of ruins, thought to be of Atlantean times, have

been found in northern Matto Grosso and in the Amazon, supporting the hypothesis the Atlanteans had colonized the area. As in Asia and Tibet, large caverns are found in Brazil with the largest being in Brazil.

The most famous of the South American tunnels is called the "Roadway of the Incas," which stretches for several hundred miles south of Lima, Peru and passes under Cuzco, Tiahuanaco, and the Three Peaks, then proceeds to the Atacambo Desert. Another branch opens to Arica, Chile.

Legends claim that the Incas had used these tunnels to escape from the Spanish conquerors and the Inquisition. Entire Indian armies entered the tunnel and carried their gold and treasures on the backs of llamas. These tunnels were constructed by the race which had constructed Tiahuanaco long before the Incas appeared on the scene. An unknown form of artificial lighting illuminated the tunnels.

Legends concerning the early gods suggest they had come from the Inner Earth of Agartha. This includes Quetzalcoatl, who is considered the greatest of the Agarthan teachers in America and was the greatest prophet of the Mayans and Aztecs. He was tall, white, and bearded and believed by some to be of Atlantean origin. Quetzalcoatl means "feathered serpent" and "teacher of wisdom." Some legends say Quetzalcoatl arrived in an aerial vehicle, left the same way, and returned to the subterranean world from where he came. He was said to be of a good appearance and wearing a flowing white garment.

Other gods that may have come from Agartha include Osiris, who was a ruler on Atlantis before Egypt, according to legend. In the Indian epic *Ramayana*, the god Rama is described as being an emissary from Agartha, who came on an aerial vehicle.

The People of Agartha

Agartha is a civilization where there is no aging or death, and it is very spiritually evolved. In this society, everyone looks young, even those who are thousands of years in age.

In Agartha, marriage does not exist, and the sexes live apart. Each sex is free and independent, and one sex does not depend on the other for economic support.

Reproduction is by parthenogenesis, which is reproduction

by development of an unfertilized female gamete that is found in plants and invertebrate animals. The children are virgin born. They are raised collectively by special teachers and not by private families. Both child and mother are supported by the community. Agartha is a matriarchal civilization where the female is considered the normal, perfect, and superior sex.

Agarthans possess superior scientific technology of which flying saucers are but one good example. They can travel to other star systems. Their great intellect results from their superior brains which are highly energetic. This is because their vital energies flow up to their brain rather then being dissipated through the sexual channels. Sex indulgence is not part of their lives. Because of their vegetarian diet, their endocrine systems are in a state of perfect balance and harmony. Their great concern for the environment prevents the health issues that are associated with metabolic toxins. Being toxin free, the people live in complete continence, and conserve all vital energies which are then converted into superior brain power.

Bulwer Lytton writes about Agartha in his book *The Coming Race*. He describes them as a subterranean civilization far in advance of our own. They are connected to the surface by tunnels. The immense cavity inside the Earth is illuminated by a strange light not produced by lamps. He also confirms that they are all vegetarians, free from disease, and have a perfect organization, in that each receives what he or she needs without exploitation.

Russian explorer Ferdinand Ossendowski was told by the lama that the tunnels which encircle the earth and pass under both the Atlantic and Pacific were built by a preglacier Hyperborean civilization. They flourished in the polar region when it was still tropical. The Hyperboreans were considered a race of supermen who possessed a high degree of scientific power that included marvelous inventions. One invention was a tunnel-boring machine that honeycombed the Earth with tunnels, but we know nothing about it.

TELOS

Many witnesses hiking on Mount Shasta have encountered unique individuals who have told the hikers they live within the mountain and are descendants of a lost continent in the Pacific Ocean named Lemuria. One such person is Princeton Winton, who experienced several strangers on Mount Shasta who told him that they were from Telos located under Mount Shasta, and their civilization originated 12,000 years ago from Lemuria. Most of the inhabitants who live under the Earth's surface are so evolved that they can communicate telepathically. Some of these advanced beings have been able to communicate with a woman named Diane Roberts who has written several books about the information including one called *Telos*.

Because the information fits in so nicely with the nontelepathic sources, I think it is important that we discuss it because of its confirmation. The two main telepathic sources for Diane are Adama who is an Ascended Master and High Priest of Telos, and Mikos of Catharia, who is linked to the Library of Porthologos, the library where all the Earth's records are kept in the Inner Earth.

According to the High Priest Adama, Telos is a Lemurian colony under Mount Shasta in northern California. He calls it a City of Light that is governed by a Council of 12 Ascended Masters and himself. Telos means "communication with spirit." Adama told Diane that the inhabitants of Telos can astral project and telecommunicate with anyone in any locality of time and space. The governing city is within the Inner Earth and is called Shamballa, confirming many other sources. Within the earth's crust are 120 subterranean cities; collectively these subterranean Cities of Light are Agartha.

The Inner Earth beings are highly evolved, and many are ascended souls who have chosen to continue their evolution there because of the perfect conditions. Their physical appearance is much like ours, and they live in the Third Dimension. Diane was told that the people from Atlantis and Lemuria are flourishing in their underground cities. They live in peace and prosperity beneath the Earth. Before a thermonuclear war took place over 12,000 years ago, the surface people were told to evacuate to the underground. About 25,000 souls with roots to Lemuria were able to do so, with the remaining perishing on the surface.

The underground people live hundreds and even thousands of years in the same body. Adama is 600 plus years old. They attribute this to their mass consciousness which holds thoughts of only immortality and perfect health. Adama said their isolation from negative extraterrestrials has allowed them to rapidly evolve. They are also conscious of harmful toxins and take great precaution in eliminating any toxins.

Physical Characteristics

The people from Telos are taller in stature and broader than surface humans. Even though they are vegetarians, they are hefty and strong. Their diet has slowed their aging process to the point where they have stopped getting older. Because souls are immortal, they can live in the same body for as long as they choose. Ascension occurs when the soul wants to move on. The people of Telos take great pride in their physical fitness, and they work out physically similar to surface humans.

They also have the ability to heal all imaginable illnesses and can replace severed limbs and organs. To do this, they work primarily with the etheric body to restore lost limbs and body parts.

Lifestyle

People from Telos begin their day in meditation and prayer. They also rest, relax, and socialize on Sundays. Because the weather is always perfect, they love picnics and eating outside. In Telos, sharing is an integral part of the civilization, and they share all.

Telos people are one big extended family who strive to be loving and tolerant. They have reached the level of unconditional love through practice from having lived many lives. In Telos, the people are always celebrating, especially with families and friends. After meals, they like to socialize and often they sing and dance together until dusk. They go to sleep early and rise with the dawn.

The key to this successful civilization is the purity of their thoughts and love for life. One can describe their lives as being joyous. Much of their life is spent outdoors and in community. Because of their lifestyle, they are not stressed, and they try to create their dreams during their waking hours.

Government and Economy

Telos is governed by a Council of 12 Ascended Masters, which is their number of completion. Councils are convened on a regular basis. Conflicts are resolved individually and according to universal law. There is no profit at the expense of another. Telos has no tax system. Food and commodities are freely given and bartered for. All goods have equal value, and things are priced within a certain range so they may be exchanged any number of times. Life is easy in Telos, and provides all they need in half the work time, allowing plenty of time for creativity.

Buildings

Homes in Telos are round in shape and made of a certain crystal-like stone that emits light, allowing people to see at all angles and in all directions. The homes are formed from a substance that allows people to look out but not allow others to see inside. The people of Telos who like to socialize are also very private.

They have special equipment that can align the home and building to the Earth's magnetic grids, as well.

Food

The people of Telos are vegetarians, eating only those foods that contain the life force within them. Their diet consists only of vegetables, grains, fruits, and nuts. All meat is banned.

Everyone in Telos works in the hydroponic gardens growing their food. When the food is harvested, they take it to their distribution center where everyone comes to pick up their supplies.

All food is eaten fresh, preserving all the nutrients. They never freeze their food because the people pick up their food supply daily. Because the workday is only four hours, they have much time available to them for exercise, cooking nourishing foods, creativity, and spirituality. Life is not stressful.

Everything is recycled over and over again, and they have no need for landfills. No paper or plastics are used, and food is free of chemicals and contaminants. Containers are used over and over again. Because they use hemp, trees are not cut down for paper.

Climate

The weather forecast in Telos is always perfect, balmy and cool. They are of the belief that on the surface of the Earth, our chaotic thoughts affect the weather. In Telos, the atmosphere is protected by the light of their thought, always in harmony with God and Earth. The air is constantly clean and clear, and there is no pollution. At one time they had an opening to the surface, but the air pollution in Telos was increasing, so the system was closed.

The vegetation is always green in Telos, with a spring-like atmosphere all year round, and flowers bloom constantly. The temperature is always in the low 70s. Light clothing is worn throughout the year.

Transportation

Tunnels intertwine throughout the planet connecting every large city and state. Destinations can be reached within hours, if not minutes. Passageways also connect the subsurface people to the Inner Earth. The people of Telos enjoy traveling to the Inner Earth where they enjoy the ocean and mountains on their vacations. It only takes them a few hours to make that journey. Electomagnetic and crystal energy are used to navigate around the inside of the globe, and all energy is free.

Tunnels to the surface of the earth are both camouflaged and guarded. The openings at the poles are camouflaged by cloud coverage to prevent the military from locating the entrances. A magnetic force field around the polar openings also helps provide camouflage.

On Mount Shasta there are many openings to the surface. The message from Adama is that when the time is right, there will be hundreds of thousands departing Telos for the surface. They will bring technology with them to the surface, and they will also be looking for homes to stay in.

Another gift the people of Telos possess is the ability to astral project. Adama said they have the ability to astral project to any location in space and time.

Evolvement

The Telos civilization, which includes about 1.5 million people, wanted to evolve in an environment without negativity.

They chose to remain silent because they have seen and experienced the negative extraterrestrials from other star systems. They are aware of the harmful effects the negative extraterrestrials have on the surface. Their experiment in life was to see how far they could evolve without war and poverty. If they came to the surface, they feared they would bring negativity back to their civilization because the surface consciousness was not ready for their teachings and technology. They have built an underground utopia where they are free to evolve.

Soon they will emerge to the surface and help the surface lightworkers with the final stages leading to the Ascension. When they emerge, they will bring with them a computer network system that links them to the rest of the universe. Telos is connected by amino acid computers to all the star systems in our galaxy. These computers are connected to the Akashic records and can convert the records into readable data.

They believe that all life needs peace for evolution, and they have created an environment of peace and prosperity. Without peace, a person struggles for survival and never has the time to add to his or her experience the strength and wisdom they have accumulated. Evolvement is a requirement for the continuation of the species.

In order for Earth and humanity to continue to ascend in consciousness, the Telos inhabitants believe the whole planet needs to be unified and merged into one light – from below the surface and above the surface.

Their emergence will be directed by the Galactic Command as to the time. This will occur after a great wave of energy (photon belt) descends upon the Earth to bring all life forms into a high level of consciousness. This is the purpose of the Mayan calendar that provides the time frame when this will happen. Once the consciousness of the surface is raised, the people from Agartha will emerge to the surface.

THE HOLLOW EARTH

Diane Roberts also received messages from the Inner Earth communicated by a highly evolved being named Mikos, whom she writes about in her book *Messages from the Hollow Earth.*

Mikos is associated with the Great Library of Porthologos, located in Catharia of the Inner Earth. The library had been linked with the Library of Alexandria and contains all the Earth's records.

Both the Catharians and people from Telos are working to help surface humans into the upcoming ascension process. Catharians chose this underground environment so they could evolve in peace and tranquility. Mikos said there are currently several million people who reside in the Hollow Earth.

Those who reside in the Hollow Earth are very advanced, having never lived on the surface but coming from other planets in our solar system and other galaxies. They are here to oversee Earth and to evolve without negative extraterrestrial influence. They have the ability to protect themselves from outside intrusion. Once humanity has evolved enough, they will emerge on the surface and help us. Besides being spiritually evolved, these Hollow Earth beings are technologically advanced beings living in peace and bliss.

The governing city within the Hollow Earth is called Shamballa, located inside the very center of the planet. It can be accessed through the openings of the North and South Poles.

Mikos said all planets are hollow. They are formed by hot gases thrown from the sun into an orbit. The shell of a planet is created by gravity and centrifugal forces, and the poles remain open. This process forms a hollow sphere that has an inner sun, smokey in color with pleasant full spectrum sunlight. The Central Sun is the mysterious power source behind the Earth's magnetic field. The environment within is highly conducive for vegetation and human life. The ratio of land mass to water is 3:1, with the interior oceans and mountains in a pristine state. The Central Sun does not move and but hangs there held by the force of gravity and is perfectly balanced, thus remaining in place.The Northern and Southern Lights are reflections from the Hollow Earth's Central Sun that emanates out of the polar openings. The sky is the very center of Hollow Earth, and there is no night, only day.

Hollow Earth remains pristine because the people don't walk or build upon the land. They travel in electromagnetic vehicles which levitate a few inches above the ground. They use free energy to light their cities, homes, and tunnels. Crystals are coupled with electromagnetic energy generated by a small sun with full spectrum lighting with a life span of one-half million years. This gives them all the power they need.

The two main portals to Hollow Earth are located at the two poles. They were closed off in the year 2000 because our government was setting off detonations at the poles to blow open entrances into their world. A magnetic force surrounds the polar openings, helping to camouflage the two entrances. Their technology has protected the openings from both the land and sea. There are also entry caverns all over the Earth where interaction can take place.

The Catharians are tall people, with the tallest being 23 feet in height, Mikos says he is 15 feet tall. Catharians also live on the planet Jupiter. Mikos told Diane there are 36,000 humans from the surface that live inside the Earth. However, during the past 200 years, only 50 surface humans have gone inside the Earth to live and only eight within the past 20 years.

The Library of Porthologus

The history of all dimensions is shared, stored, and preserved in the Library of Porthologus. Nothing is lost, and every life that has existed is recorded and has been woven into the meaning of the universe's tapestry. The library is located below the Aegean Sea. In the past, there were surface entrances to the Library of Porthologus with one entrance being the Library of Alexandria that was destroyed in 642 A.D.

Mikos said he is very old and a resident of the city of Catharia. He serves as a library researcher compiling records of both Earth and the Universe. The library has the technology to enable one to read about historical events and actually experience them firsthand from the crystals that store the memories of all events. Mikos is considered an ambassador for the whole Earth, and he journeys to other star systems to arrange for their records to be transferred to Porthologus for safe keeping. The library houses all the records of the universe. It is also multidimensional with an interdimensional portal that can take you to where ever you project your thoughts.

Mikos said that someday humans will be able to travel to the library through the tunnel system that opens between the two civilizations. This will occur when humans stop warring and reach unity consciousness. Once that is achieved, the Inner Earth civilizations will open to us and we will experience the joy and freedom of being in unity. We will benefit from all their wisdom and accomplishments. Mikos said that one day we will open our eyes, and we will all have unity consciousness.

The Library of Porthologus stores all its records on Telonium plates, an ancient and eternal kind of metal that lasts forever and never shows any signs of decay. The library is vast in its size, round in shape, and is located in a huge carved-out cavity. The ticket to the library is encoded in our DNA.

Environment

Within the Hollow Earth are beautiful mountains and oceans. The height of the mountain range is in direct proportion to the dimensions of the earth cavity and towers above the landscape. The oceans flow calmly and swiftly around the inside of the globe. The oceans are large with waves that are affected by the moon. Water in the oceans and rivers is composed of living consciousness. It is this water consciousness that keeps the inhabitants young forever. People swim in the ocean for great distances and no one ever drowns.

Hollow Earth cities are nestled in lush woodlands that overflow with flowers and huge trees that are thousands of years old and adorn the landscape. Vegetation surrounds all man-made structures. There is nothing but beauty that creates this environment of bliss. The temperature in the Hollow Earth is in the low 70s.

Lifestyle

People of the Hollow Earth are strong, healthy, and robust, and they are vegetarians who do not believe in hunting. These advanced beings have freedom, health, and abundance.

Transportation is free, and they travel between the Hollow Earth and subterranean cities in a short period of time. Transportation is efficient, clean, and quick. The intertwining tunnels allow Inner Earth inhabitants to get to their destination within hours, if not minutes. These underground passageways have been used for eons, and people can travel freely without reservations.

The buildings in Inner Earth are round and translucent with unique properties. Their shape allows the energy to move freely and revolve, helping to keep the dwellings clean. Mikos believes that in the future, surface homes will be round. Inner Earth people live in great palaces that are infused with light. Their homes are set in the lushness of the countryside blending into the landscape with many surrounded by lakes and streams. They have no cities like those on the surface but only country. Their grand palaces are made of crystallized stones embedded with jewels from the earth. The crystallized stone creates a magnetic field that nurtures and balances their bodies. It fills them with the life force that emanates from the great Central Sun.

Inner Earth people are very tall, and the width and girth of their figures are more than double ours. The size of their fruits and vegetables are also huge, providing a healthy diet for these vegetarians, as all their foods are organic. When purchasing goods they barter because there is no monetary system, similar to Telos.

Bodies of Hollow Earth people do not age and buildings do not deteriorate. There is no such thing as aging, because they never see it. Their existence is in a state of divine perfection with a timeless environment. They never hurry and are never late. No thoughts are less than divine.

They dress according to how they feel, often in loose, bright, soft fabric clothing made from hemp and other vegetable material. Trees are never used for anything except for the beauty of nature. Often the people walk barefooted on earthen paths lined with flowers, as there are no streets.

Neither disease nor pollution exist in Hollow Earth. They believe that all diseases on the surface are somehow associated with toxins and greed. They never get tired, sick, angry or worry. Because of their mindset and lack of pollution, they can live hundreds and even thousands of years.

Their workday is less than half of ours on the surface. This gives them time to balance mind, body, and creativity. Culture is very important to them as they enjoy music, dance, and theater, and creativity fills their lives. Theaters are located everywhere. Sharing is the key to their life, not owning.

Communication is by telepathy. Only when one is in a harmonious mood can telepathic communication take place. Mikos emphasizes that we are all telepathic and can converse with him or other beings in Hollow Earth.

Mikos told Diane that time is different on the Earth's surface compared to the Inner Earth. People need to repeat life lessons over and over again on the surface until they are learned and mastered. When a planet evolves and is in an ascended state, there is no more time because from the higher perspective of consciousness, one can see into eternity. One then experiences multidimensionality and can experience all states of consciousness. All is one, and all is simultaneous. Time speeds up when you are wiser and have grown in consciousness. Time is an illusion, claims Mikos.

Hollow Earth has a spaceport that spreads out for hundreds of miles and is directly aligned with the openings at the North and South Pole. The portals are wide enough for some of the motherships to enter. None of their space-crafts have crashed because every component of the craft is monitored by amino acid computers which can detect and correct any problem immediately.

For eons, the beings of Inner Earth have been calling for more light and help from the Confederation of Planets. According to Mikos, they want the Confederation to intercede and stop the "influx of ravaging bands of extraterrestrials who have been scouring space to find planets like Earth that are rich in resources. Surface beings are direct descendants of roving bands of extraterrestrials who created us to mine Earth's resources."

Hollow Earth people are a good example of what a society can be without war and poverty. Our civilization on the surface has the opportunity to be happy, healthy, and fulfilled through a proper mindset and healthy environment. Surface humans will eventually awaken to experience life as it should be, the example set by the civilizations of Inner Earth.

Chapter Ten

UNDERGROUND BASES
Out of Sight, Out of Mind

Throughout the book we have alluded to secret underground bases, those by the extraterrestrials, those by the government, and those by a cooperative effort between the Secret Government and extraterrestrials. These bases should be highly secret, but some secrets cannot be kept. Researcher Valerian Valdman states that he knows of innumerable witnesses who have confirmed the existence of underground bases that the Grays use for their genetic manipulation experiments, mostly on abductees. Both military and civilians confirm this information. Former MI-6 intelligence officer James Casbolt tells of 4,000 underground bases that the intelligence agencies know about.

Richard Sauder, Ph.D., is a financial and military researcher in the Southwest who has conducted extensive research on underground bases. He has concluded, "I consider it an absolute certainty that the military has constructed secret underground facilities in the United States, above and beyond approximately one dozen known underground facilities." He reports some of the known sites are located at Fort Belvoir, Virginia; West Point, New York ; Twentynine Palms Marine Corps Base, California; Groom Lake, Nevada; White Sands Missle Range, New Mexico; Table Mountain near Boulder, Colorado; Mount Blackmore and Pipestone Pass, Montana. Sauders claims, "There are actual tunneling machines that crawl though the ground like giant mechanical earthworms with huge appetites."

Dulce
One underground base that houses both the Grays and the military is located near Dulce, New Mexico. Arthur Horn, Ph.D., writes about Thomas C. who was working in a joint U.S. Government and Gray alien underground project under

Dulce. Thomas took a number of black and white photographs, created a video tape, and acquired printed material regarding the underground base. He made five copies of the material that later became known as the Dulce Papers. Thomas had witnessed row after row of humans and human remains in storage. As a consequence of his indiscretion, Thomas lost both his son and wife to experiments by the Grays.

The underground facility at Dulce is connected to other facilities by tunnels that are used by underground high speed trains.

The Grays (the government calls them EBEs - Extraterrestrial Biological Entities) have a genetic disorder of their digestive system. To sustain themselves, they need to use an enzyme or hormonal secretion from tissue extract acquired from humans and animals. The secretions obtained are mixed with hydrogen peroxide and applied to the skin by dipping parts of their body in solution. The body absorbs the solution and then excretes the waste back through the skin. This was the purpose of cattle mutilations in the United States from 1973 - 1983 and some human abductions. Various parts of the bodies were taken to various underground laboratories, one of which was the Dulce facility.

The Dulce facility is described as large and jointly occupied by the CIA and aliens. It has huge tiled walls that seem to go on forever. Witnesses have seen huge vats with amber liquid containing parts of human bodies stored inside. George Andrews writes, "The aliens seem to absorb atoms to eat. The DNA in cattle and humans is being altered. . . . Some humans are kidnapped and used completely. Some are kept alive in large tubes and kept alive in amber liquid. Some humans are brainwashed and used to distort the truth."

James Bishop has written a book entitled *The Dulce Base* that contains descriptions of the interior of the base from people who have worked there. The facility has multiple levels, great security procedures, and uniformed personnel. It posts a list of its board of directors' names who are in high positions of public government.

Level Six is described as "Nightmare Hall," as it holds the genetic laboratories. Inside sources say bizarre experimentation has resulted in multi legged humans, reptilian humans, winged humanoids, Gargoyle-like beings, and Draco Reptoids.

Going down to Level Seven, witnesses say, is even worse. There are rows of thousands of humans and humanoid mixtures in cold storage. Some humans are in cages, usually dazed or drugged. Some cry out for help. In 1978, a small group of workers discovered the truth, which began the Dulce War. It had become apparent that some of the nation's missing children had been abducted by the negative Grays and their body parts utilized for experimentation. Many ended up at Dulce. A special armed forces unit was used to try and free a number of people held captive in Dulce's underground base. Called the Dulce War, 66 soldiers were killed and none of the people were freed.

Valdman Valerian, in his book *Matrix II*, writes extensively about the Grays, and Dr. Arthur Horn believes that few people have more information about aliens than does Valerian. According to Valerian, the Grays are spreading end-time predictions as part of psychological control over humans. He presents evidence that the Grays are interfacing with humans in secret societies and within government complexes. He claims the Grays and their allies have been controlling and manipulating humans for thousands of years. The extraterrestrials and the Secret Government want to create fear, because it is their tool for controlling humanity.

A *Courageous Scientist*

Not many people would risk their life to disseminate truth, but Phil Schneider made the supreme sacrifice to alert humanity what was happening between the Secret Government and aliens. Schneider was an accomplished geologist/engineer who helped construct deep underground military bases (DUMB) and worked for such companies as Morrison Knutson that contracted with the U.S. Government and NATO. After working on the secret bases for 17 years, he decided to go public by giving 30 public lectures and surviving 13 assassination attempts, giving his last lecture in November 1995 in Denver, Colorado before being killed the following January.

When Schneider gave his 1995 talk, 131 U.S. underground military bases and 39 underground prisons had been constructed up to depths of two miles. He personally worked on 13 underground bases including the Dulce, New Mexico base where the alien-human battle transpired causing 66 deaths and putting Schneider in the hospital for over a year with radiation burns. He also worked at Groom Lake (Area 51) for 11 years. All the 131 U.S. bases are connected by tunnels, often located by nearby military bases. The tunnels were made by advanced laser technology supplied by aliens that allowed seven miles of tunnels to be built daily. The laser would pulverize rocks into a fine powder that was then congealed like an agate substance lining the tunnels. The tunnels can measure up to 28 feet by 28 feet housing tracks that allow subterranean trains to travel at Mach 2 and roads for traditional vehicle traffic. In the lower 48 states, there are 127 bases with three of them located near Denver. One of them has eight levels located below Denver International Airport; another in Colorado Springs (talked about in this chapter), and another near Fort Collins. (This is the second source stating Fort Collins as a site, as I thought they may be confusing it with the military base of Fort Carson, CO.) There are 9 underground bases around Nellis AFB in Nevada. An enormous underground base was being built in Sweden in 1995, costing nearly $2 trillion according to Schneider.

The astounding yearly budget of over $500 billion a year comes from the Black Operation Budget approved by Congress, from illegal drug activities of the CIA, and clandestine operations of the National Security Agency. The United Nations is also involved. Each base costs between $17 - 26 billion.

Schneider said the secret military technology supplied by aliens is about 45 times more advanced than present day technology, and it would take 1,000 years for civilian technology to catch up. For decades, alien supplied technology has been able to control weather with the ability to stop and start hurricanes, and Russia has worked with the U.S. in weather modification since 1972.

Schneider said there are eleven alien civilizations routinely involved with our planet, with two of them truly benevolent. He says the aliens are working closely with the New World Order (Secret Government) who believe they are doing it for the higher good but are being duped by the negative extraterrestrials, who want to take control over our planet and cause mass depopulation.

Cheyenne Mountain

Another underground base that is jointly run by the Secret Government and aliens is located within Cheyenne Mountain, Colorado, near Colorado Springs, the same mountain that houses NORAD. The entrance to this facility is on a different side of the mountain than NORAD's entrance. It reportedly has five levels with reports of negative Sirians, negative Orions, and negative Grays being involved in the operation. A number of reports tell of children being taken in and out of this underground base by bus. There is some evidence that there may be a Satanic connection.

Groom Lake

The most famous of the underground bases is located in Area 51 at Groom Lake, Nevada, an "above top secret base." Robert Lazar made Area 51 famous when he went public on a Las Vegas television program telling about his experience at Area 51. Lazar, a physicist who was employed by the government, worked at Area 51 to reverse engineer the propulsion system of a crashed UFO that had been retrieved. The purpose of reverse engineering is to gather as much information as possible from this advanced technology. Lazar was worried about his personal safety and life after he quit working at Area 51. Going public, he felt, would be insurance for his safety because the government is in denial that such a base exists.

Between 1972 and 1974, a huge undergound facility was constructed with the help of extraterrestrials at Area 51. The government had bargained for technology at Groom Lake, but ironically the technology developed could only be operated by the extraterrestrials.

Carlsbad

Researcher Jason Bishop writes of the Blondes having a base in the Carlsbad Cavern area of New Mexico. The Blondes had refused to give our government weapon technology and tried to warn the government about entering into an alliance with the Grays. Upon our government's entering into its alliance with the Grays, the Grays requested that our government detonate nuclear weapons underground in the Carlsbad area to destroy the Blondes' underground base, and our government complied. The Blondes abandoned the base. The area in which the "underground testing" occurred is being used as a storage area for nuclear waste.

Pine Gap

According to George Andrews' research, the United States has three major underground bases in Australia. Pine Gap is the top-secret base that is completely financed by the United States government and is officially called a "Joint Defense Space Research Facility." This base is used mainly to study electromagnetic propulsion. Pine Gap is about five miles deep and is likely used as an underground antenna to recharge the batteries of submarines in the Pacific or Indian Oceans through ELF broadcasting. Pine Gap is well known as our most important control center for spy satellites. There are rumors that the Grays are stationed at Pine Gap and three other underground bases in the southern hemisphere – Guam, South Africa, and Australia.

Kirkland AFB

Information given to Alex Collier by the Andromedan Moraney told of a facility under Kirkland AFB in Albuquerque, N.M. with an entrance in the Monzoni Mountains. The facility was operated by the National Security Agency (NSA). The Andromedan said that the facility builds free energy devices for use in space and on the moon and Mars by the United States (The Secret Government has bases on both the moon and Mars). The underground facility is 29,000 square feet also houses a Secret Government jail for captured extraterrestrials.

Moraney listed a number of corporations assisting the aliens and Secret Government which included: Standard Oil, Lockheed, Northrup, McDonald Douglas, AT&T, IT&T, Utah Mining Company and others. It is interesting to note that the NSA is exempt from all U.S. laws.

John Lear

John Lear is considered one of the most respected authorities in the UFO/extraterrestrial field and is the son of William Lear who invented the Lear jet. Lear was a test pilot for the Department of Defense and also worked for the CIA. He has taken personal risks by divulging top secret knowledge and believes the public has the right to know. On December 28, 1987, Lear made the following statement, "The United States government has been in business with the little Gray extraterrestrials for about 20 years. In its effort to protect democracy, our government sold us to the aliens. Those that made the deal had the best of intentions."

On April 30, 1964, communication between the aliens and U.S. government took place at Holloman AFB in New Mexico. At that time, three saucers landed in a prearranged area where a meeting was held between intelligence officers of the U.S. government and the aliens. Between 1969 and 1971, representatives of the U.S. government (MJ-12) made an agreement with creatures called EBEs. MJ-12 is composed of 12 individuals and was established by President Truman to deal with UFO/extraterrestrial information. In exchange for technology from the aliens, the U.S. government agreed to ignore the abduction of humans and suppress information on cattle mutilations. The U.S. government insisted that a current list of abductees be submitted on a periodic basis to MJ-12 and to the National Security Council.

The government realized that the negative extraterrestrials had reneged on the agreement and a conflict arose on the MJ-12 committee. Some members wanted to go public with knowledge about extraterrestrials and some wanted to keep it suppressed. Members of MJ-12 included Dr. Edward Teller and Henry Kissinger and possibly Admiral Poindexter. The majority of MJ-12 voted to keep knowledge of extraterrestrials suppressed.They decided it was absolutely essential that no one, not the Senate, the Congress, or citizens of the United States, become aware of the real circumstances surrounding the UFO coverup.

John Lear gave a lecture to the Dallas Mufon group on August 10, 1988. He told the group that the nation had been brainwashed by CIA mind control operations based on fear and ridicule. Since 1947, there have been 1,000,000 abductions, and in the last 13 years 14,000 cattle mutilations have been reported. He said there are approximately 70 alien civilizations visiting Earth at the present time. He estimates there are about 10 million Grays in underground bases on Earth and on the moon. They enter underground bases through inter-dimensional transference, which is a hyperspace maneuver. He told the audience that Eisenhower had allowed the reign of power to pass from the President to the Pentagon, which he felt was a big mistake.

THE MONTAUK PROJECT

Preston Nichols is an electrical engineer who wrote about an extraordinary experience that he had at Montauk regarding an experiment in time that involved the Grays, Sirians, Orions, and the Secret Government. His book is entitled *The Montauk Project: Experiment in Time*. It all began in the 1940s when the extraterrestrials shared technology with the United States military about cloaking ships so they would not be vulnerable to attack in WWII. The experiment, subsequently called the Philadelphia Experiment, involved the ship *U.S.S. Eldridge* docked in the Philadelphia naval shipyard, whose code name was the Rainbow Project. A book was written and a movie made regarding the Philadelphia Experiment.

The experiment involved a technique that made a ship invisible to enemy radar and the naked eye. During the experiment, the ship was removed from the space-time continuum and suddenly reappeared in Norfolk, Virginia, hundreds of miles away from Philadelphia. From a physical standpoint, the experiment was a success, but to the sailors on board, it was devastating. When the *U.S.S. Eldridge* returned to the Philadelphia Navy Yard, some of the sailors were planted into the bulkhead of the ship itself. Those who survived were disorientated and mentally unfit. The project was deemed too risky, and Dr. John vonNeuman who headed the experiment was reassigned to work on the Manhattan Project to develop the atomic bomb. The Philadelphia Experiment, which took place on August 12, 1943, ripped a hole in our time/space fabric that allows UFOs to enter our time/space, which explains why there have been so many UFO sightings since WWII.

Following the war, the Rainbow Project resumed in the late 1940s under name of The Phoenix Project, headquartered out of

Brookhaven Labs on Long Island and headed again by Dr. John vonNeuman. Research found that those sailors involved in the original experiment were literally removed from space and our universe as we know it. This accounted for the invisibility of the ship and people aboard. The alternate reality that was created had no time reference at all because it is not part of the normal time stream. Essentially, it is entirely out of time.

The research needed to solve the problem of bringing human beings into the "bottle (alternative reality) and eventually out again," while at the same time connecting them to their real time reference, writes Nichols, who participated in the research. This meant that, when they were in the alternate reality or "bottle," they would have to be supplied with something that would give them a time reference. Somehow the researchers needed to feed all the natural background of the Earth into the alternate reality.

Preston writes, "The computer had to generate an electromagnetic background (or phony stage) that the physical being would synchronize with as well. If that wasn't done, the spirit and physical and nonphysical both would go out of synch, resulting in insanity. The time reference would lock in the spirit and the electromagnetic background would lock in the body. The whole project was started in 1948 and was developed in 1967."

A final report was submitted to Congress for funding, but Congress said no. They were afraid that if the wrong people obtained this technology, they themselves could lose their minds and be controlled. The project was disbanded in 1969.

The Brookhaven group was determined to continue the project that was denied by Congress. They decided to work with the military and use private money, and so Montauk AFB, a former radar base at the tip of Long Island, was reopened. Preston Nichol had heard that the original money came from Nazi gold that was recovered in WWII and also from the infamous Krupp family who controlled the ITT corporation. During WWI and WWII the Krupps were owners of German munitions factories. The research staff at Montauk was a mixture of military employees, government employees, and personnel from various corporations.

The research discovered that through microwave technology, which changed the pulse rate and width of microwave emissions, they were able to change the way people thought. They could make people cry, laugh, sleep, and become agitated. They also found that they could place thought patterns in individuals. Programs were devised that could change the moods of people, increase the crime rate, and agitate people.

In the 1950s, ITT had developed sensor technology that could literally display what a person was thinking. No one knows how the technology worked, but it was suggested that the research was aided by the Sirians, who provided the basic design. When the Montauk researchers heard about the discovery, they wanted to turn the mind-reading machine into a transmitter, which would transmit an alternative reality to the crew. After a year of research, the transmitter at Monatuk would transmit a clear representation of what a psychic was thinking. The psychic they used in the research was named Duncan Cameron, who had been on the *U.S.S. Eldridge* during the Philadelphia Experiment. By 1977, the transmitter was reproducing thought forms to a high degree of fidelity without glitches. Duncan was used because he was needed to rectify the time/space anomaly he experienced with the Philadelphia Experiment.

The experiment evolved to the point that when the psychic Cameron concentrated on a solid object, it would precipitate out of the ethers and appear somewhere on the base. The team had actually discovered pure creation that came out of thought with use of the transmitter. Whatever Duncan could think up would appear. Sometimes it would be visible but not solid to the touch. When Duncan concentrated on a person, 99 percent of the time a subject would get thoughts similar to Duncan's. In fact, Duncan could control another person and make the individual do anything Duncan wanted him to do, sometimes things he would not ordinarily do. This became the start of the mind control aspect of the Montauk project.

Another group of extraterrestrials, the Orions, gave the research group technology of an antennae called an Orion Delta T Antennae that could warp time. Duncan would be directed to concentrate on an opening in time, for example from 1980 to 1990. At this time, a hole or portal would appear right in the center of the Delta T Antennae and one could walk through the portal into the period from 1980 to 1990. It all depended on Duncan's ability to concentrate. One might go through the portal and come out in 1960. There was always a risk of getting lost in space and time. If Duncan's mind drifted, the portal would drift.

The scientists found there was no particular problem in creating a time warp, but predicting what it would do was difficult. The researchers used the 1943, 1963, and 1983 vortex bases, the natural 20-year biorhythms of the Earth. These years serve as anchor points for the master vortex with August 12, 1983, being the main anchor point. Another factor was to stabilize a spatial aspect so they could place a portal not only at a particular time but also in a particular space. The whole system was tuned to Duncan. At this stage of the research, security was tight and the team operated on a need-to-know basis. They did not want the military to know.

The research team began to experiment with time travel. Through the vortex, they could sample air, terrain, and everything without entering the portal. Those who entered the portal found that when they began to walk down it, they would suddenly be pulled through it and propelled out the other end, usually in another place anywhere in the universe. The tunnel resembled a corkscrew. Nichols writes, "On the other end you would meet somebody or do something. You could complete your mission and return. The tunnel would open for you and you would come back to where you came from. If power was lost, you would be lost in time or abandoned somewhere in the vortex. Many were lost but not deliberately or carelessly."

Nichols writes that it was routine to create a tunnel, grab someone off the street, and send them down the vortex. Often people used in the experiment were derelicts or winos whose absence would not create a fuss. If they were fortunate and returned, they would make a full report. Nichols said that many didn't make it and still are floating around in time. Many would be wired with a video camera and record where they were, with the project ending up in an extensive library of video tapes. Sometimes the researchers would use kids who sometimes did not return.

They gathered a lot of data, but the Montauk researchers wanted to shut down the project following an episode in 1983. (The movie "Total Recall" is based on some of the events that happened at the Montauk Project.) Countless missions were run until August 12, 1983, when the actual lock was made back to 1943 and 1963. On that day the *U.S.S. Eldridge* appeared through the portal. Duncan from 1943 and his brother (who had also been on the *U.S.S. Eldridge*) appeared and could be seen through the time portal. Duncan of 1983 was prevented from seeing the 1943 Duncan. Things began to go wrong as the generators on the U.S.S. Eldridge in 1943 locked onto the ones at Montauk of 1983 and created a free energy mode which prohibited them from shutting down the power, and some died. Any surviving crew members were brought to Montauk in 1983. This included personnel who were considered to be reincarnated since the Philadelphia Experiment. Duncan and Al Bielek (Duncan's brother in 1943) were both there and were two of the primary witnesses.

After this harrowing experience, the Montauk Project was shut down. If this episode is true as Preston Nichols describes it, one can understand how extraterrestrials can control minds of world leaders, the Secret Government, and humanity. It also provides evidence that they work with the Secret Government for reasons that are not always best for humanity.

Chapter Eleven

THE SECRET GOVERNMENT
Controlling the Planet

Throughout the book, I have alluded to the Secret Government, suggesting that it is the Secret Government that really runs the world. They control the economy, politics, wars, technology, environment, media, drug trade, and even religion. How can this be, as many of us live in democratic countries? This chapter will try to sort out the methods by which a few people can control humanity. Dr. John Coleman, a former intelligence officer who wrote a book entitled *Conspirator's Hierarchy: The Story of the Committee of 300,* explains the powers that control and manage the British and U.S. governments. Dr. Coleman details how the United States and Britain are controlled by 300 people, and how their methodology controls politics, economy, and most social conditions. Much of the money to run the Secret Government comes from illegal drugs. They also are working with negative extraterrestrials who have an agenda of their own and control the Secret Government. Jim Marrs has compiled probably the best overall synopsis of the secret groups that control our world in his book *Rule by Secrecy.* Information from these two books is referred to extensively in this chapter.

To put things in perspective, while writing this book, I received a document on the internet written by a former British MI-6 intelligence officer named James Casbolt who has his own website. It reads: "James Casbolt is a former MI-6 agent who worked in 'Black Ops' drug trafficking in London between 1995 and 1996. He comes from a line of intelligence people . . . . James wishes to make amends for his part in these operations and blow the whistle on the crimes against humanity that the intelligence agencies are involved in. MI-6 and the CIA have cornered the global drug trade (which is worth at least 500 billion pounds a year, more than the global oil trade) and are now bringing the majority of illegal street drugs into America and Britain.

"They are using this drug money to fund projects classified 'above top secret' which includes the building and maintaining of deep underground military bases. There are now over 4,000 of these bases worldwide and the average depth of these bases is four-and-a-quarter miles below ground level. Some are shallower and some are deeper. The bases are on average the size of a medium sized city and yes, he says, there are aliens in them.

"James is connected to ex-intelligence people who have worked in these underground bases and on other ET-related projects. There are vast numbers of children and adults disappearing around the world and ending up in these underground bases. As a former MI-6 agent, he has seen aliens firsthand and has inside information."

As we have mentioned earlier in the book, the extraterrestrials have the ability to control people, especially those in power positions.

THE INFRASTRUCTURE

Coleman's inside investigation found that the world is mainly controlled by a group of 300 individuals who make up an organization called The Committee of 300, an open conspiracy against God and man that has enslaved the majority of humans on Earth. The upper level Secret Government places itself in view of the White House, Congress, and House of Parliament. Queen Elizabeth II is the head of the Committee of 300, with individuals such as George H. W. Bush and Henry Kissinger playing major roles in the Committee. Other names given to this group are Illuminati, Round Table, and Milner Group. Coleman asserts that the final objective of the Committee is to overturn the U.S. Constitution and merge this country with a godless New World Order. The Committee of 300 took its present form in 1897, but has been in existence for over165 years.

The Club of Rome, established in 1968, is the foreign policy arm of the Committee of 300, with Henry Kissinger carrying out their policy when he was Secretary of State. Coleman claims that Kissinger played a major role in destabilizing the U.S. by means of three wars – Korean, Vietnam, and Gulf Wars. The long-range plan of the Committee is to destabilize the Middle East, which is taking place today through the Iraq War.

Colman asserts the Committee of 300 is the ultimate controlling body that runs the world and has done it for over a century through secret societies, front organizations, government agencies, banks, insurance companies, international business, and the oil industry. For example, NATO's policies were formulated by the Club of Rome, according to Coleman.

Coleman emphasizes that the Club of Rome is a conspiratorial umbrella organization and is a union between Anglo-American financiers and Old Black Nobility families of Europe. It has its own private intelligence agency, uses Interpol, all U.S. intelligence agencies, the KGB, and Mossad. Coleman explains that Britain has controlled the U.S. through the Committee of 300. Some of the goals of the Committee of 300, as gathered from documents by Coleman, include the development of a One World Order; destruction of national identity; mind control; eliminating nuclear power; depopulating large cities; legalizing drugs and pornography; eliminating 3 billion people through wars, starvation, and diseases; weakening the moral fiber of workers; collapsing the economy; subverting governments from within; organizing a world terrorist apparatus; controlling American education and destroying it; and collapsing the American steel, auto, and housing industries. These are but just a few goals, and, as we can see, many of these goals are beginning to manifest.

The Round Table

The Round Table was founded by an Englishman named Cecil Rhodes (1863 - 1902), who originated the Rhodes Scholarships to promote feelings of universal citizenship. His fortune was created through gold and diamond mines in South Africa. Rhodes was instrumental in making South Africa a vital part of the British Empire. His goal was to create a one world government led by Britain and to make English the universal language. Rhodes wanted to accomplish his goals through a Brotherhood network (the Round Table) that became corrupted, an institution that he created that fell into the hands of those who wanted to oppress the human race. The Round Table ended up creating the concentration camps and the atomic bomb, the things and principles that Rhodes had dedicated his life to preventing. The Round Table was a secret society patterned after Freemasonry. His chief supporter was the English banker Lord Rothschild. Following Rhodes' death in 1902, the Round Table gained support from leaders of the international banking community such as the Rockefellers, Whitney, J.P. Morgan, and Carnegie.

Following WWI, several subgroups evolved from the Round Table and became very powerful societies in England and the United States. Lionel Curtis established a local chapter in England called the Royal Institute of International Affairs (RIIA), and in the United States, the Round Table evolved into the Council of Foreign Relations (CFR).

The CFR is considered a think tank for foreign policy, and its members have dominated presidential administrations. David Rockefeller was chairman for many years. The banking industry exercised a strong influence on American politics, especially in foreign affairs via the CFR. This has resulted in a status quo for inflation, debt, and war.

Royal Institute for International Affairs (RIIA)

Dr. Coleman claims that nothing in the United States can happen without the sanctions of the Royal Institute for International Affairs. The Committee of 300 issues its orders through fronts like the RIIA. Through a 1938 agreement, the United States intelligence agencies are obligated to share intelligence secrets with British intelligence. The RIIA has control over U.S. politics, writes Coleman. For example, everyone who runs for president of the United States is selected by the CFR, acting on instructions from the RIIA. Following the 1980 election, all key policy making positions were filled by CFR nominees.

RIIA established the largest brainwashing facility in the world called the Tavistock Institute for Human Relations as part of Sussex University. Tavistock became the nucleus of Britain's

Psychological Warfare Bureau. It designed methods to get the United States involved in WWII. During WWII, Kurt Lewin from Tavistock established the OSS, the forerunner to the CIA. Lewin was director of the Strategic Bombing Survey, which was a plan to bomb German worker housing instead of military targets such as munitions plants. The munitions plants belonged to the international bankers who did not want their assets destroyed. Their purpose was to demoralize the workers.

Tavistock programs also led to establishing the Office of Naval Intelligence (ONI), which is the number one intelligence agency in the United States, dwarfing the CIA in size.

Contracts worth billions of dollars were given to Tavistock to run over 30 research institutions. One way the Committee of 300 influences humanity is through opinion polls. David Yankelovich is a Committee member and one of the most respected of all pollsters. He asserts that polling is a tool to change public opinion. New public opinion on almost any subject can be created and disseminated around the world in a matter of two weeks. David Naisbett in his book *Trend Report* was commissioned by the Committee of 300 to write a book that describes all of the techniques used by public opinion makers to bring about the public opinion desired by the Committee of 300.

Coleman claims that the governments of Britain and the United States have the machinery to bring us in line with a New World Order. It has established control networks that are binding. One of the tools they use is fear, such as terrorism and communism during the Cold War. He believes we are being brainwashed to give up Constitutional rights and accept every lawless act carried out by the government, almost without question. A good example of this policy is action taken after 9/11 and the war in Iraq. There is also good evidence that the Secret Government was behind 9/11 as revealed by David Icke in his book *Alice in Wonderland and The World Trade Center Disaster*, and Jim Marrs book, *The Terror Conspiracy*. Coleman argues that the greatest dangers facing free people today are coming from Washington D.C.

Coleman is correct when he says the vast majority of

Americans can see no motivation for what has been going on and can't buy into the conspiracy. Until motivation is shown, all information is rejected. Besides greed and power, one has to look at the negative extraterrestrial motivational factor, which is to keep us in conflict and trapped within the Third Dimension. They are the ones who control the Secret Government.

Stanford Research Institute

In 1946, Tavistock Institute for Human Relations founded the Stanford Research Institute (SRI). It is the largest military think tank in the United States and includes specialty departments for chemical and biological war experiments. According to Dr. Coleman, SRI contracted a top secret program for the Pentagon to develop special bombs that could trigger volcanoes and earthquakes. Coleman had few kind words for Professor Willis Harmon, Ph.D., who was the director of the SRI Center for the Study of Social Policies. Harmon later became president of the Institute of Noetic Sciences. Harmon's report for SRI was about "Changing Images of Man." People have succumbed to what Tavistock calls "future shocks" meaning that we are so numbed by cultural shock that nothing surprises us any more.

Council on Foreign Relations

The Council on Foreign Relations (CFR) is considered the father of the modern American secret societies, which began as an outgrowth of meetings conducted during WWI. In 1917, Edward Mandell House, President Woodrow Wilson's confidential advisor, had gathered about 100 prominent men to discuss the postwar world. They made plans for a peace settlement, which eventually evolved into Wilson's famous "fourteen points." The "fourteen points" were global in nature and called for removing all economic barriers, equality of trade conditions, and formation of a general association of nations. Both British and American delegates met in Paris to form an Institute of International Affairs, with one branch in the United States and one in England. The English branch became the Royal Institute of International Affairs, which was to guide public opinion toward accepting one world government or globalization.

The U.S. branch was established on July 21, 1921 as the Council on Foreign Relations (CFR). Article II of the new bylaws stated that anyone revealing details of CFR meetings in contravention of the CFR's rules could lose membership. CFR is officially a secret society. Since 1945, the CFR headquarters is housed in the elegant Pratt House of New York City, which had been donated by the Pratt family of Rockefeller's Standard Oil. Jim Marrs writes that many of its members belong to the upper crust Social Register groups. The invitation only membership includes more than 3,300 leaders in finance, commerce, communications, and academia. Admission is very difficult and discriminating. Early funding for the CFR came from bankers and financiers such as John Rockefeller, J. P. Morgan, and Jacob Schiff. Today, funding comes from major corporations such as Xerox, General Motors, Texaco, Ford Foundation, Andrew W. Mellon Foundation, Rockefeller brothers, and many more.

Critics claim that the CFR has its hands in every major 20th Century conflict. Many authors believe that CFR is a group of men set on world domination through multinational business, international treaties, and world government. Admiral Chester Ward, a longtime CFR member, is quoted as saying "CFR, as such, does not write the platform of both political parties or select their respective presidential candidates, or control U.S. defense and foreign policy, but CFR members, as individuals, acting in concert with CFR members, do." Admiral Ward said that one common objective of CFR members is "to bring about the surrender of the sovereignty and the national independence of the United States. . . . Primarily, they want the world banking monopoly from whatever power to end up in the control of global government." In Jim Marrs' book *Rule by Secrecy*, he quotes Phyliss Schafly in her book *Kissinger on the Couch*, "Once the ruling members of the CFR have decided that the U.S. Government should adopt a particular policy, the very substantial research facilities of CFR are put to work to develop arguments, intellectual and emotional, to support the new policy, and to confound and discredit, intellectually and politically, any exposition." John Kenneth Galbraith resigned from the CFR in 1970 and wrote about their meeting, "Why should businessmen be briefed by Government officials on information not available to the public, especially since it can be financially advantageous?"

Until 1988, there have been 14 U.S. Secretaries of State, 14 Treasury Secretaries, 11 Defense Secretaries, and scores of other federal department heads that have been CFR members. Nearly every CIA director has been a CFR member. CFR members who take government positions tend to bring in fellow members. The Clinton administration was top-heavy with 100 CFR members at the beginning of his administration. There are about a dozen members of both the House and Senate who are CFR members. It needs to be emphasized that very few people in these positions really know the big picture and are compartmentalized, believing they are doing the right thing for humanity. They are being manipulated.

Gary Allen in his book *None Dare Call It Conspiracy* writes, "There really was not a dime's worth of difference between (presidential candidates). Voters were given the choice between CFR world government advocate Nixon and CFR world government advocate Humphrey. Only the rhetoric was changed to fool the public." Allen believes that both Democrats and Republicans must break the insider control of their respective parties. Both Al Gore and George W. Bush have long standing business and family ties to Wall Street and CFR members.

For nearly a century, the CFR has had a powerful influence over U.S. policy. This influence is shared by other secret societies such as the Trilateral Commission and the Bilderbergers.

Trilateral Commission

One of the secret groups most talked about by the conspiracy groups is the Trilateral Commission. Zbigniew Brzeginski brought the concept of the Trilateral Commission to David Rockefeller when Brzezinski was head of the Russian Studies Department at Columbia University. When Brzezinski was at the Brookings Institute, he researched the need for closer cooperation between the trilateral nations of Europe, North America, and Asia. In July of 1972, the Trilateral Commission began organizing with the blessing of the Council of Foreign Relations and the Bilderbergers, and it was officially founded on July 1, 1973.

Headquarters are located in New York, Paris, and Tokyo. Many of the members of the Trilateral Commission are in positions of power and have acquired a reputation for being the Secret Government of the West. The Trilateral Commission has been described by some as a cabal of powerful men out to control the world by creating a super national community dominated by multinational corporations.

Many politicians have been members of the Trilateral Commission, including Jimmy Carter, Walter Mondale, and much of the Carter cabinet. Republicans include Henry Kissinger, Casper Weinberger, and George H. W. Bush. Paul Volcker and Alan Greenspan, both former heads of the Federal Reserve Board, were members of the Trilateral Commission.

Senator Barry Goldwater was concerned about the Trilateral Commission and wrote, "What the Trilaterals truly intend is the creation of a worldwide economic power superior to the political government of the nation states involved. As managers and creators of the systems, they will rule the world." Secretary of State for President Carter, Edmund Muskie, charged that Brzezinski, head of the National Security Agency, was making foreign policy rather than coordinating it. Congressman Larry McDonald, who was national chairman of the John Birch Society, was critical of the secret societies such as the Trilateral Commission, Bilderbergers, and CFR. He died in the controversial downing of the Korean Airline 007 on September 1, 1983. Ronald Reagan criticized both Jimmy Carter and George Bush for belonging to the Trilateral Commission. Following his election, Reagan's 59-member transition team consisted of 28 CFR members, 10 members of the Bilderberger Group, and 10 members who were part of the Trilateralists.

Jim Marrs quotes Tex Marrs (no relation), president of Living Truth Publishers in Austin, Texas, "The Trilateral Commission is a group with the goal of hastening the era of World Government and promoting the international economy controlled behind the scenes by the Secret Brother (the Illuminati)."

The Bilderberger Group

The Bilderberger group consists mainly of European Royalty who meet in secret each year to discuss the issues of the day.

Conspiracy theorists believe they conspire to manufacture and manage world events. Bilderbergers often belong to the CFR and Trilateral Commission as well. Creation of the Bilderberger group came in the early 1950s. Members include prominent businessmen, politicians, bankers, educators, media owners, and military leaders. Bilderbergers are closely tied to European Royalty from Britain, Sweden, Holland, and Spain. Dutch Prince Bernhard Pietier was the primary force to begin the Bilderberger group. He was a former member of the Nazi's SS and is a major shareholder of Dutch Shell Oil. After WWII, banking magnate Victor Rothschild and Polish socialist Joseph Retinger encouraged him to create the Bilderberger group.

Dr. John Coleman writes, "The Bilderbeger Conference is a creation of (Britian) MI-6 under the direction of the Royal Institute of International Affairs." James Marrs argues that the annual Bilderberger conference is partially organized and sponsored by the CIA. Investigative reporter James P. Tucker writes, "The Bilderberger agenda is much the same as its brother group, the Trilateral Commission – the two groups have an interlocking leadership and a common vision of the world. David Rockerfeller founded the Trilaterals but shares power in the older Bilderberger group with the Rothschilds of Britain and Europe." The Bilderberger group meets annually at a plush resort in total secrecy with about 120 attending. However, there is evidence that its recommendations often become official policy. For example, the idea of a common European currency was discussed by this group before it became policy. They also have political clout. Sources say they take credit for putting Bill Clinton in office and ousting Margaret Thatcher.

The Rockefeller family in America and Rothschild family in Europe are two of the most powerful families in the world. Rockefellers made their money in oil and banking while the Rothschilds made their fortune mainly in banking. Another major player in the world was businessman and banker J. P. Morgan. Author David Icke claims that both Morgan and Rockefeller used Rothschild financing to "build vast empires which controlled banking, business, oil, steel, etc., running the United States economy the way the Oppenheimers do in South Africa."

The Federal Reserve System

Those who control the money control the world, and this is where the Federal Reserve system comes into play. Jim Marrs writes, " The ultimate control of money rests with the bankers of the Federal Reserve." The Federal Reserve system consists of twelve Federal Reserve banks, each one serving a section of the country but mainly dominated by the New York Federal Reserve Bank. The Federal Reserve banks are administered by a board of governors appointed by the President and confirmed by the Senate. The power wielded by the Federal Reserve is enormous, as any change in interest rate is reflected in the stock market and business world.

"The real story of the Fed is who controls it," writes Marrs. "Using a central bank to create alternate periods of inflation and deflation, and thus whipsawing the public for vast profits, has been worked out by the international bankers to an exact science."

Eustace Mullins writes in her 1983 book *The Secrets of the Federal Reserve*, "An examination of the major stockholders of the New York City banks show clearly that a few families, related by blood, marriage, or business interests still control the New York City banks, which in turn hold the controlling stock of the Federal Reserve Bank of New York." The private bank of the New York Federal Reserve is owned by Chase Manhatten Bank (a Rockefeller bank), which owns 32.35 %, and Citibank which owns 20.56%, thus controling the majority.

The major banks finally obtained a longstanding goal – taxpayer liability for the losses of private banks. Paul Warburg, a mastermind financier, is quoted as saying that "Federal Reserve notes constitute privately issued money with the taxpayers standing by to cover the potential losses of the banks which issued it." A 1963 Federal Reserve publication states, "The function of the Federal Reserve is to foster a flow of money and credit that will facilitate orderly economic growth, a stable dollar, and long run balance in our international payments."

Marrs writes in his book quoting William Greider, "Most Americans have no real understanding of the operation of the international money lenders. . . The bankers want it that way. We recognize in a hazy sort of way that the Rothschilds and Warburgs of Europe and the houses of J. P. Morgan, Kuhn, Loeb and Company, Schift, Lehman, and Rockefeller possess and control vast wealth. How they acquire this vast financial power and employ it is a mystery to most of us." He continues, "International bankers make money by extending credit to government. The greater the debt of the political state, the larger the interest returned to the lender. The national banks of Europe are actually owned and controlled by private interests." Private interest groups also own the national banks in both Europe and the United States.

Marrs wrote, "Today, America is experiencing an inflating depression – prices continue to rise because of an inflated money supply. The more money that is in circulation, the less it is worth." William Bramley argues, "The result of this whole system is massive debt at every level of society today. The banks are in debt to the depositors, and the depository money is loaned out and creates indebtedness to the borrowers. Making this system even more akin to something out of a maniac's delirium is the fact that banks, like other lenders, have the right to seize physical property if its paper money is not repaid." This is the main concern of conspiracy theorists who believe that the Secret Government is planning to crash the economy and to acquire control of humanity by calling in the assets because the debt cannot be paid.

ILLUMINATI

When conspiracy theorists talk about the Secret Government and those who run the world, the term Illuminati is often used as a generic term to encompass those in control. Who are the Illuminati? The concept of Illuminati can be traced back in history to early sects who possessed esoteric knowledge. It was in May of 1776 when the Illuminati order was first publicly identified. The Bavarian Illuminati was formed by Adam Weishaupt, a professor of Canon Law at Ingolstadt University of Bavaria, Germany. A cofounder was William of Hesse, an employer of Mayer Rothschild. Rothschild and German Royalty were connected through Freemaonry. Weishaupt had studied the esoteric knowledge of Egypt and Persia as well as an ancient form of Gnosticism. Weishaupt spent several years developing a plan to consolidate all occult systems into the new Illuminati order. He created a pyramid structure of degrees for his initiates based on the Jesuit and Freemason structure, with key personnel located within the top nine degrees. Marrs writes about a 1969 magazine article that claimed the Illuminati originated within a Muslim Ismaili sect that was connected to the Knights Templar, who may have brought Illuminati ideals to Europe, centuries before Weishaupt. According to the article, Weishaupt studied the teachings of the infamous Muslim Assassins who achieved "illumination," by ingesting homegrown marijuana. Essentially, the Illuminati philosophy was based on esoteric knowledge, and they used it to oppose the Catholic Church and the national government the Church supported. Weishaupt believed man was corrupt because of religion, the state, and bad examples that perverted him. He also believed that the end justifies the means – meaning lies, deceit, theft, murder, or war could be used to achieve their goals. The key to Illuminati control was secrecy. Weishaupt never wanted the Illuminati to appear anywhere in its own name, but always covered by another name and another occupation. Freemasonry was used as a cover for the Illuminati, similar to the Thule Society which seeded Nazism. Weishaupt set out to deceive the public so he reminded Illuminati's top leaders to hide their true interactions from their own initiates. He is quoted "speaking sometimes in one way, sometimes in another, so that one's real purpose should remain impenetrable to one's inferior." Weishaupt's followers were enlisted by the most subtle methods of deception and led on toward a goal entirely unknown to them, according to John Robison, a knowledgeable Freemason.

Weishaupt and the Illuminati wanted a world government based on their philosophy of human-centered rationalism, with the world government being administered by themselves. Weishaupt said, "The pupils (of the Illuminati) are convinced that the order will rule the world. Every member therefore becomes a ruler."

In 1777, Weishaupt blended his Illuminati with Freemasonry when he joined the Masonic Order Lodge Theodore of Good Counsel in Munich. In 1783, the Bavarian government saw the

Illuminati as a direct threat to the establishment and barred the organization. Many members fled Germany taking Illuminatis with them and thus spreading it. Weishaupt had created a pyramid chain of command, making it difficult to detect who ran the organization.

The headquarters of Illuminized Freemasonry was moved to Frankfurt, Germany's center of banking controlled by Rothschild. For the first time, Jews were now being admitted into the order. The Frankfurt Lodge included Rothschild, who was Jewish. The Illuminati were now safely hidden within Freemasonry, and the conspiracy against the Church and monarchy could be hidden. From the Frankfurt Lodge, the great plan of world revolution was carried out.

Jim Marrs writes, "Many researchers today believe the Illuminati still exist and that the order's goals are nothing less than the abolition of all government, private property, inheritance, nationalism, the family unit, and organized religion. Certainly the goals and methods of operation still exist. Whether the name Illuminati still exists is really irrelevant."

Some of these Illuminati philosophies have been carried forward within the Masonic lodges. The Masons have played an important role in both the American and French Revolutions. Jim Marrs writes, "Conspiracy researchers make it clear that Illuminized Freemasons have used any and every opportunity to advance their cause, regardless of which side they may support at the moment. . . . The Masonic slogan *ordo ab chao* generally is regarded as referring to the order's attempt to bring an order of knowledge to the chaos of the various human beliefs and philosophies in the world – a New World Order." Writer Tex Marrs interprets *ordo ab chao* as a secret doctrine of the Illuminati. He states, "They work to invent chaos, to generate anger and frustration on the part of humans and thus, take advantage of people's desperate need for order."

Whether Illuminati is the proper term that should be used for the Secret Government is debatable but the principles underlying the two groups seem to be similar. The next question that needs to be asked is, where does the Secret Government get its money?

THE DRUG TRADE

Governments need money to operate, and the Secret Government is no different. Congress authorizes some money for Black Operations that are highly secretive, and most Congressmen don't know the specifics. Black Operations are secret projects that supposedly enhance our national security. According to former intelligence officers John Coleman and James Casbolt and other conspiracy researchers, drug trade is how the Secret Government is financed, and drug trafficking is controlled from the top down by the Committee of 300.

Drug trade was started in Britain with the British East India Company (BEIC) in the 19th Century followed by the Dutch East India Company, both controlled by the Committee of 300, according to Coleman. BEIC was responsible for addicting China to opium. They established the "China Inland Mission," whose purpose was to have the Chinese people become addicted to opium that then created the market which BEIC filled. A certain percentage of England's budget depended on proceeds from drug trafficking, which Parliament did not want to abolish in the 19th Century. John Coleman writes, "The history of British occupation of India and Britain's opium trade was some of the most dastardly blots on Western civilization." Approximately 13 percent of income in India, under British rule, was from the sale of good quality Bengal opium distributed to China by BEIC. The British created the opium market in China and then filled it. The British Royal family joined the BEIC drug trade, used it to produce opium in Bengal and elsewhere in India, and then controlled exports through transit duties. Prior to 1896, the trade of opium was still illegal but there was no attempt to stop drug trafficking because the profits were so enormous and benefitted England so richly. In fact, the BEIC tried to sell opium to the Union and Confederacy of the United States in pill form as a pain killer.

Coleman writes, "The BEIC and British governments held the monopoly in opium trading, and the only people to make good fortunes were the nobility, aristocracy, plutocrats, and oligarchical families of England. . . . many of whose descendants sit on the Committee of 300." Since 1729, every British monarch has benefitted from drug trade, including the present monarchs who still profit from it. The British opium wars against China were fought to put the Chinese in their place. The resultant Hong Kong agreement solidified the Chinese-British partnership, which established an equal partnership in opium trade. Hong Kong is the base of the lucrative opium trade. Before the rapid industrial development in China, Coleman asserts that the Chinese economy was tied to the economy of Hong Kong and would have taken a terrible beating if it were not for the opium trade.

Heroin trade is a big business that is run from the top down by some of the most untouchable families in the world, with each family having a member on the Committee of 300. Indian opium profits went directly into the Royal coffers and the pockets of nobility to make them billionaires. Coleman asserts that trafficking illicit drugs is the largest enterprise in the world. Heroin trade is financed by the London Bank, Hong Kong banks, and some Middle East banks. Lebanon is becoming the Switzerland of the Middle East, with drug money being deposited in secret bank accounts. The Bank of International Settlement and the International Money Fund are banks that are clearing houses for drug trade, writes Coleman. Opium trade would not function without the banks. International Money Fund agents are quoted at a seminar that "they could literally cause a run on any country's currency using narcotic dollars, which would precipitate a flight of capital." Coleman claims there is not a single government that does not know precisely what is going on with regard to the drug trade.

Coleman's findings found that America is run by 300 families and England is run by 100 families. These families are connected through matrimony, business, financial institutions, and have ties to the Black Nobility of Europe, the Freemasons, and the order of St. Johns of Jerusalem. Surrogates act on behalf of the families to protect huge shipments of heroin.

There are pockets within the intelligence agencies that are involved in drug trafficking, including the CIA in America and MI-6 in England. During the Vietnam war, opium was grown in the Golden Triangle of Thailand, Burma, and Laos. There were a number of reports that heroin was being shipped back to the

United States in body bags and coffins on military transport planes. Colonel Bo Gritz discovered this drug trafficking in Vietnam and spent many years of his life trying to awaken America to what was going on. James Marrs wrote in *Rule by Secrecy* that the December, 1988 bombing of Pam Am Flight 103 over Lockerbie, Scotland was perhaps connected to CIA drug smuggling. The Pam Am flight victims included a CIA team, which was on its way back to Washington to report its discovery of CIA drug smuggling and drug running activities in the Middle East with financing through a Masonic Lodge in Italy.

Coleman is of the belief that the Committee of 300 has taken full control of the South American cocaine trade. He wrote that "Bush was literally pushed into invading Panama and kidnapping Noriega who had become a serious impediment to (drug) trade in Panama. The CIA and U.S. Army intelligence are on record as giving Noriega $320,000. Following the invasion, nothing was found to implicate Noriega as a drug dealer."

Poppies were the major crops of Afghanistan before the Taliban took control of the country. They prohibited the farmers from growing poppies, thus seriously affecting the drug market. After the U.S. invasion of Afghanistan, Afghanistan is again the world's leading producer of poppies. Conspiracy researchers wonder if opium may have been a major motive in the U.S. invasion that was orchestrated by the Secret Government. Other major producers of opium are Iran, Pakistan, and Lebanon. The major route of heroin into Europe is through Monaco. Here it is transported across the open border to France into French laboratories. Coleman claims Princess Grace was killed in a car accident because Prince Ranier, ruler of Monaco, got greedy.

John Coleman reports that he was able to access a top secret RIIA paper. It reads, ". . . Those who have been without jobs for five years or more will turn way from the church and seek solace in drugs. That is when full control of the drug trade must be completed in order that the government of all countries who are under our jurisdiction will have a monopoly, which we control through supply." He asserts that the CIA and MI-6 of England have played a major role in the drug trade with the goal of creating the monopoly as described above.

Coleman writes, "The Committee of 300 is the ultimate

secret society. The opium trade is still run by the same upper class families of Britain and the United States. This is the real reason why the drug problem is not eradicated."

We now can explain where the Secret Government acquires its money, and as James Casbolt stated in the beginning of the chapter, it is worth 500 billion pounds a year. This money enables the Secret Government to construct underground bases and maintain a liaison with the negative extraterrestrials, who have an agenda of their own.

Chapter Twelve

THE EXTRATERRESTRIAL INFLUENCE
Gifts and Curses from the Heavens

As we shall see in this chapter, almost every aspect of human life has been influenced by extraterrestrial civilizations from science to agriculture to religion to wars. The rapid advancement of scientific knowledge can be attributed to the extraterrestrials. They also have the ability to influence wars and distort religions. How can this be when only a few individuals acknowledge the existence of extraterrestrials? Extraterrestrials have the ability to influence thought, they possess the most advanced technology, and they have knowledge about spiritual truths. Most of them, while adhering to a code of noninterference, are here to help us evolve out of the Third Dimension, whereas several of the negative extraterrestrial groups have shared secrets with the Secret Government to advance their own agendas. As you will see, we are being manipulated by the extraterrestrials!

Agriculture

To catch a historical glimpse of the extraterrestrials' influence, we need to look at the ancient cuneiform writings of Sumer in Mesopotamia. The writings tell how the gods (Anunnaki) gave humans the knowledge and capability of farming and animal husbandry, laying the foundation of civilization.

However, following the Great Flood about 12,000 years ago, the planet Earth was devastated. The Sumerian writings tell about the Anunnaki providing grains and tuber vegetables to humans to replace the food sources destroyed in the Great Flood. All the domesticated plants had been destroyed in the Deluge,

221

but the Anunnaki had foreseen the consequences and sent seed to their planet Nibiru for storage prior to the Flood. Once the waters settled down, the god Enlil sent the seeds back to Earth. The writings say that it was the Anunnaki who had introduced wheat, barley, and other cereal grains to Sumer, even prior to the Flood. Following the Flood, the lowlands were uninhabitable because of the wetness, and this resulted in the introduction of agriculture in the highlands of the Near East. It was Enlil who taught humans how to terrace mountains and plant seed. Enlil went to "the mountains of aromatic cedar," believed to be the mountains in Lebanon, where he taught the people the art and science of mountain terracing. The god Ninurta was revered by the Sumerians, partly because he taught humankind how to farm by introducing the plow to them.

Livestock were also introduced to Earth by the Anunnaki. They were responsible for bringing into existence "wooly creatures" (sheep) that multiplied, and they also introduced domesticated cattle.

Sumerian scholar Zecharia Sitchin believes the Sumerian god Enki and the Egyptian god Ptah were one and the same. No matter, it was Enki who was responsible for draining and cleaning the Nile Valley and preparing it for the great Egyptian civilization. In Egyptian lore, it was Ptah who was given credit for reclaiming Egypt from inundating water. In Central and South America, it was the gods such as Quetzalcoatl and Varichoca that provided knowledge about agriculture.

Modern Technology

Toward the end of the 20th Century, a remarkable book by Philip Corso, a retired Army intelligence officer, appeared on the scene. The book, *The Day After Roswell*, describes the after effects and scientific evolution that followed the UFO crash in Roswell, N.M., with the information coming from an inside source, Col. Philip Corso. The crash took place on the night of July 1, 1947, northwest of Roswell, New Mexico. Several Indian artifact hunters saw the craft go down and reported it to the sheriff's office. Just before the civilians arrived, the Army was finishing its investigation, as it had been tracking the craft on radar and realized it had gone down. The craft was mostly intact, dark in color, of a delta shape with two fins on the top side of the delta feet. On board the craft, the Army investigators discovered five little gray beings measuring about 4.5 feet tall. One of them was still alive and tried to escape, but was shot.

Dan Dwyer, a civilian fireman who arrived on the scene, saw small bodies being lifted onto stretchers and then into a transport truck, one still alive. Dwyer described the beings as having an "oversized balloon-shaped head, the size of a child with humanlike features. It was grayish brown and completely hairless." Dwyer picked up a gray metallic cloth that belonged to the craft and rolled it into a ball and released it. The metallic fabric snapped back into shape. He kept the cloth, not disclosing its existence to the military.

In 1944, Philip Corso was in a top-level position in Italy. He had been ordered to oversee the formation of a civilian government under Allied Military rule. Following the war, Corso enrolled in Military Intelligence School at Fort Riley, Kansas. This is where he had seen one of the alien bodies. He described it as having a six-fingered hand, no thumbs, thin legs and feet, a tiny nose that didn't protrude, and a flat slit for a mouth. The body was without hair, no cheeks, and large almond shaped eyes with no pupils. It was a sight he would never forget.

During the 1950s, Corso was on the National Security staff and worked under President Eisenhower. In 1961, Col.Corso was assigned to the Pentgon to work under Lt. General Arthur Trudeau, with whom he had worked before. Corso was to head up the Foreign Technology department in Army Research and Development.

In the Pentagon, there was a filing cabinet that included the Roswell material, and Trudeau asked Corso to go though the cabinet and file a report about its contents. Included in the cabinet were artifacts from Roswell. Trudeau was very fearful of the Secret Government members within the government who had been placed in the Pentagon by the spy master of the Kremlin. Because of the distrust within the military branches and intelligence systems, General Trudeau told Corso that he was to bypass the military and get the Roswell material to defense contractors to do research and development on the alien artifacts. Only Corso and Trudeau knew of the plan to bypass the military. Besides, the Army, Air Force and Navy each had its own cache from Roswell. The Air Force had kept the craft, and Corso believed they also had the bodies. According to Corso, the Navy was having its own problem with USOs, unidentifiable submerged objects. During the 1950s, whatever was flying at thousands of miles an hour in the air would dive into the ocean and navigate there just as easily. Corso conjectured that the USOs were building bases on the ocean bottom.

It was 1959 when America's top rocket scientist, Wernher von Braun, had gone on record by announcing that the military had acquired new technology as a result of top secret research on UFOs. An elite secret Air Force unit that operated out of Fort Belvoir, Virginia, supposedly an Army base, was responsible for retrieving downed UFOs. Fort Belvoir became a repository of classified UFO information, including film. Astronaut Gordon Cooper had described film that he saw at Edwards AFB in 1957 of a UFO landing at the base. Joe Walker, an X-15 pilot, said he filmed UFOs during an X-15 flight in 1962.

Colonel Corso said the government had established a secret project they called Horizon, which was to be a defense fortification on the moon to be ready to fight against a Soviet attempt to use the moon as a military base. In 1961, NASA agreed to serve as a cover with military planners to work as a second tier space program, which in reality served as a coverup within the civilian scientific mission. Corso also wrote that extraterrestrial craft were buzzing the lunar module on successive missions after they were thrust out of Earth's orbit. The Army and Air Force accumulated at least 122 photographs taken by astronauts on the moon that showed the presence of aliens. This is the reason that the Reagan administration pushed so hard for the Space Defense Initiative in 1981. There was a dual purpose strategy during the Cold War to survey both extraterrestrial activity and Soviet activity. Instead of using Cape Canaveral for their secret mission to the moon in the Horizon Project, the Army chose Brazil. There is also evidence that we have a base on Mars, according to an expose' documentary called Alternative Three, where a number of British scientists mysteriously disappeared for a period of time; some have not returned. The Secret Government has long held alien scientific technology to accomplish such missions as the Mars base in cooperation with the negative extraterrestrials.

General Trudeau and Lt. Col. Corso also feared that the Secret Government would bury the technology and keep it for themselves. They devised a plan to release the technology into the civilian sector. Trudeau wanted to filter the Roswell technology into the mainstream of industrial development through the military defense contractors. This was to be Corso's mission. He developed a list of technological items from the downed space-craft that he wanted researched. Besides government contractors, universities were utilized by Trudeau and Corso to research and develop this new technology which included:

1. Image Intensifiers – Night Vision Devices
2. Fiber Optics
3. Super Tenacity Fiber – Kevlar
4. Lasers
5. Molecular Aligned Metallic Alloys
6. Integrated Circuits
7. Project Horizon – Moon Base
8. Portable Atomic Generators – Ion Propulsion Drive
9. Irradiated Food
10. Particle Beams – Antimissile Energy Weapons
11. Electromagnetic Propulsion Systems
12. Depleted Uranium Projectiles

As a result of this hidden agenda to research the Roswell UFO artifacts by the civilian sector, lasers, fiber optics, accelerated particle beam devices, and bullet-proof Kevlar material were discovered – all from a crashed alien space-craft. Corso believed these discoveries paralleled the technology the Nazis had in WWII, also obtained from extraterrestrials. Stealth technology was also developed as a result of the ET space-craft. Other military technology that arose from the crash were guided missiles, the antimissile missile, and "the guided missile launched accelerated particle beam firing satellite killers."

From a crashed UFO, scientific technology resulted that changed the world forever. It was the courage of General Trudeau and Colonel Corso who made sure it was available to humanity, rather than being suppressed by the Secret Government.

Religion

The extraterrestrials have had a great influence on religion, mainly for control reasons. It is easy to understand how an extraterrestrial event could be misunderstood for a sacred sign from God. Extraterrestrials have been with us from the beginning of humanity. William Bramley, in his book the *Gods of Eden*, argues that most religions are custodial religions influenced by the negative extraterrestrials in order to control humanity.

Judaism and Christianity

Extraterrestrials are the gods of the Bible. Zecharia Sitchin believes that Genesis is mostly an edited Hebrew copy of the ancient Mesopotamian texts, the history of the Anunnaki. Following Genesis, the Old Testament tells of the violence and brutality that the Lord God Jehovah generated on the enemies of the Hebrews and sometimes on the Hebrews themselves. Most of the gods of ancient history were extraterrestrials who used their superior technology and metaphysical knowledge to deceive and control humans. Jehovah, the Old Testament God, was one of these extraterrestrials. Former college professor at Colorado State University, Arthur Horn, in his book *ET Influence on Humankind,* believes that Jehovah was a captain of a space-craft, and angels were crew members of a UFO. Jehovah was not an etheric being, but was often directly involved in day-to-day lives of Hebrews. He could be ruthless. Dr. Horn believes "that all of the major and minor religions of the world have either been founded or distorted or influenced by different groups of extraterrestrials." This includes the New Testament and Christianity, where the virgin birth, the angel Gabrielle's visit to Mother Mary, and the Star of Bethlehem are, in all probability, ET related.

Hinduism

In the Hindu religion, many of the gods of ancient Hindu writings are depicted in the skies as riding in remarkable and powerful chariots. They fire tremendous weapons from these chariots, which can emit powerful beams of light. Erich von Daniken and other writers believe the gods of the Vedas and other Hindu writings are extraterrestrials, and their conflicts deal with other extraterrestrial groups of the world. Horn believes that humankind has been spiritually inhibited through false religions propagated by negative extraterrestrial groups. For example, Hindus are taught that they are born into the caste system of their fathers where they will remain throughout their lifetimes. If the Hindu is good and obedient to the ruling caste, he might be able to reincarnate into a higher caste in the next lifetime. The caste system becomes a measure of one's spiritual evolution in Hinduism. This justifies what kind of treatment lower castes receive from the higher caste. Today, the caste system is against the law.

Buddhism

In the Buddhist religion, author William Bramley believes the concept of Nirvana as taught by most Buddhists is quite distorted. Buddhist teachings emphasize that all physical reality is bound up in pain. The purpose of a Buddhist on the path is to escape physical reality and to move to a state of blissfulness and nonexistence or Nirvana, which has been translated as the void or nothingness.

Buddha had taught that a true state of Nirvana " . . . originally referred to that state of existence in which the spirit has achieved full awareness of itself as a spiritual being and is no longer experiencing suffering due to misidentification with the material universe." Bramley writes that a spiritual entity would have a difficult time evolving past a certain stage if it did not remain conscious of its surroundings and strove instead for a state of nonexistence.

Islam

Bramley argues that Islam is another custodial religion inspired and influenced by negative extraterrestrials for the purpose of dividing and controlling humankind. He believes the angel who communicated with Mohammed was orchestrated by the negative extraterrestrials making Mohammed believe he was receiving the true religion. Perhaps, though, the spiritual truth of the angel was subsequently distorted. There are two primary divisions of Islam – the Sunnites and the Shiites. The Sunnites are more numerous and practice a form of Islam believed to be closer to what Mohammed taught than what the Shiites practice. The Shiites have splintered into numerous sects over the centuries, including the infamous Assassins who fought the Christian Crusaders. Today, while the Iraq war is being fought, one can see the deep division between the Sunnis and Shiites.

Both Bramley and Horn conclude from their research that all religion contains spiritual distortion and corruption. All major belief systems are distortions of the truth of our existence, secretly promoted by the negative extraterrestrials to keep humanity divided.

Mormons

William Bramley believes the Mormon religion is another religion greatly distorted by the extraterrestrials. The seeds of Mormonism began in the spring of 1820 when Joseph Smith of Manchester, New York, experienced a paranormal event. A pillar of light appeared over his head and descended upon him when two beings stood above him and began dictating pronouncements that laid the foundation for the Church of Jesus Christ of the Latter Day Saints. One of the beings, named Moroni, made subsequent visits and gave Joseph Smith knowledge aboute future Mormon Church doctrine. One night the angel Moroni told Joseph about the existence of ancient plates regarding the history of the early North American continent. Joseph was instructed to dig up the plates and translate them to the world. The same exact apparition appeared to Joseph on another night. To Bramley, this image sounded like a UFO encounter, with the apparition being a hologram. Bramley believes this is but another example of a custodial force pretending that they are God and interfering in human affairs.

The Book of Mormon is said to be a translation of the ancient metal plates that Smith had excavated at the command of the angel Moroni. The style of writing resembles the Old Testament. According to the writings, people from Palestine were transported in saucer-like submarines to the Americas under the guidance of God in the year 600 B.C. God was sending them to a new world because of the Tower of Babel incident. The angel encouraged their human servants to practice important virtues, especially obedience.

Bramley thinks that the Book of Mormon may be one of the most significant historical records to come out of the custodial religions. The arrival of the Palestinians coincides with the date that historians have assigned to the emergence of the ancient civilization of Mexico and Central America, explaining their abrupt rise. The Book of Mormon also describes a nuclear blast, written long before nuclear weapons.

Mormonism encouraged humans to welcome the grim fate of endless entrapment in human bodies. In ancient times, like modern times, the Sumerian texts told how the gods wanted to join spiritual beings to human bodies so that the custodians would have a slave race.

During a UFO contactee experience, we know that extraterrestrials can block memory. The reason our memory is blocked, according to Mormon doctrine, is to ensure that our choice of good or evil would reflect our earthly desires and will, rather than the remembered influence of our good Heavenly Father. Bramley thinks that the passage suggests there exists a custodial intention to block human remembrance of a Supreme Being.

Mormon doctrine states that wars continue to occur over the generations as God's toll for maintaining control. Bramley argues the use of breeding wars is a custodial toll for maintaining control over the human population.

Keeping with the tradition of custodial religion, Mormons were told, as were other religions, that "their religion was the only true and living church upon the face of the whole Earth." This keeps in flux a division among religions to maintain people in conflict.

The Mormon church has the world's largest geneological library, housed inside a mountain 20 miles south of Salt Lake City. The collection of statistics has produced over 60,000 rolls of microfilm recording deaths, marriage licenses, births, cemetery lists, etc. Mormons are taught that knowing family lines is necessary in order to trace all those who lived and died in the past so they can be blessed. Geneological records are kept for the entire human race, as these records are important to Mormons. In early Mormon philosophy, Aryanism was an important aspect, but Mormons have now dropped this racial belief. Several writers have suggested that Mormons are connected to the Illuminati and claim that the geneological records help the negative extraterrestrials monitor DNA around the world.

Joseph Smith claimed that he patterned the Mormon priesthood according to the dictate of an angel. Smith had also become a Freemason for a short period of time to attain background knowledge for founding the Mormon religion. Brigham Young was also a Freemason, and he led the Mormon exodus across the country to Utah.

Spirituality

Bramley asserts that the custodial extraterrestrials have done great harm to the human species and have retarded their spiritual development. They have spread misinformation and distorted spiritual truths that often have been incorporated into the major religions. He makes a comparison between custodial religions and maverick religions, showing how the maverick religions were custodialized by the custodian extraterrestrials. For example, the maverick religions are inspired by highly spiritual teachers such as Jesus and Guatama Siddharta, the Buddha. Maverick religions emphasize the individual as the basis of evolution, not a Supreme God. Custodial religion teaches that individuals should adhere to a particular doctrine that is based on factors such as obedience. With maverick religions, spiritual salvation is entirely up to the individual, but in custodial religions salvation depends on the grace of God.

THE EXTRATERRESTRIAL INFLUENCE ON HEALTH

The Plague

Extraterrestrials not only have influence on religion and wars, but they can also influence our health, claims William Bramley. During the 14th Century, one of deadliest plagues to encompass the planet was the Black Plague. It began in Asia and soon spread to Europe, where it killed over 25 million people between 1347 and 1350. Also called the Bubonic Plague, it continued to infect Europe with decreasing fatality rates every ten to twenty years in short-lived outbreaks until the 1700s. Around 100 million people eventually died from the Plague.

One of the greatest mysteries about the Plague was how isolated human populations became infected. A great many people throughout Europe and other regions of the world reported an outbreak of the Plague following exposure to a foul smelling mist that would frequently appear following an unusually bright light in the sky. The bright light and mist was reported far more frequently than were rodent infestations. The Plague years coincided with a period of UFO activities.

In Europe, the first outbreak of Plague occurred between 1298 and 1314 when seven large comets were seen over Europe. At that time, any unusual object in the sky was called a comet. The objects were sketched by some observers showing portholes, which suggest that they were UFOs, not comets. In Asia, reports of strange flying objects spraying a mist that destroyed the fertility of the land suggested a chemical or biological defoliant.

In the year 1618, reports of eight or nine comets preceded subsequent epidemics. In 1606, a comet again seemed to precede a general plague, and there are many more similar examples, according to Bramley. Physicians in the era of the Plague took it for granted that the strange mists caused the Plague. In fact, one account stressed that the Plague did not spread from person to person, but was contracted by breathing "the deadly stinking air." Other accounts say that people actually saw the Plague coming through the streets in a pale fog.

Today, we know that the Plague can be transmitted through germ warfare, as both the United States and Russia have stockpiles of this biological weapon. Any person breathing in the mist of plague microorganisms will inhale the disease. Enough germ weapons exist to wipe out a good portion of humanity. One might conclude that the Plague of the 14th Century was probably germ contamination purposely caused by extraterrestrials to depopulate the Earth. The Bible says that plagues were one of Jehovah's methods of punishing people for evil, and we have given evidence that Jehovah was probably an extraterrestrial.

Another ramification occurred during the Plague years. False accusations circulated stating that Jews were causing the Plague by poisoning wells. Christian communities naturally became fearful and hated the Jews, and many Christians participated in the genocide of Jews. Some claim that more Jewish lives were lost then than during the 20th Century genocide of Jews. Bramley writes that the genocides were often instigated by German trade guilds, which were offshoots of ancient secret Brotherhoods. Corrupted Brotherhoods have been a major source of historical genocide, such as the Nazis.

AIDS

During the 1970s the AIDs epidemic that affected mainly homosexuals, drug users, and the poor in Africa began. Robert Strecker, M.D., Ph.D., physician/pharmacologist discovered evidence that AIDs had been a genetically engineered virus that was introduced to Africa through vaccines with the purpose of depopulating segments of the population. Strecker wrote a book about the evidence, which resulted in his brother, who also was talking about the evidence, being assassinated. Dr.Arthur Horn writes, "The ruling elite decided to target what they considered elements of society for their new plague . . . . In 1978, 1979, 1980, and 1981, the U.S. population was infected by an experimental Hepatitis B vaccine that was administered by the Center for Disease Control in the cities of New York and San Francisco." In other words, the Secret Government and extraterrestrials have the ability to cause pandemics.

William Cooper, a former member of the United States Naval Intelligence Service, had access to top-secret material that revealed the U.S. Secret Government's involvement with an extraterrestrial species. He wrote a book entitled *Behold a Pale Horse* claiming that the Secret Government originated in the 1940s following the Roswell UFO crash and was behind the AIDs epidemic. Cooper did not specifically claim that the extraterrestrials were directly involved with the AIDs epidemic, but the secret societies and organizations behind the spread of AIDs were in contact with the extraterrestrials at the highest level. An extraterrestrial source called the Pleiadian Plus, who communicates through Barbara Marciniak, also suggests that the negative extraterrestrials were behind the spread of AIDs.

WARS

Researcher William Bramley set out on an ambitious investigation to try to find the cause of war, as he writes in his book *The Gods of Eden*. His research led him to the conclusion that wars are caused by a small group of men whose purpose is to control and profit from the wars of humanity. This small group of men operate in secrecy and often use secret societies to help carry out deceit and manipulation of the human population. During his investigation, he kept discovering an extraterrestrial factor and concluded that extraterrestrials were behind most events in the world in order to keep humanity divided and ignorant. He concluded that there definitely was an ET conspiracy to manipulate humanity, and that the conspirators used war as a tool for social and political control over large populations.

Bramley concurred with Zecharia Sitchin that humans were created as a slave race and were treated harshly by the early gods of ancient history – the Anunnaki. Calling these extraterrestrials custodial gods, Bramley claims these gods have influenced human affairs for a long time. They had gone underground, both literally and figuratively, to manipulate humans by influencing certain secret societies, which most often had several levels of initiations and secrecy. The highest and most secret level of these societies had custodial gods who exacted their power and influence. For example, a few people taught some sort of discord, like racism, to members of a secret society who were at a lower level of understanding. The secret society would covertly agitate for the cause, which could lead to an armed conflict. The irony is that splinter groups of the same secret society would support the opposite side of the issue of conflict.

Bramley asserts that custodial extraterrestrials provide

humans in the highest position of these secret societies with power and control. These people are recipients of enormous monetary gains, which help finance the ensuing war. He gives an example of the religious wars occurring after the Protestant reformation and the numerous revolutions that occurred in the 18th, 19th, and 20th Centuries. Historians acknowledge that secret societies gave root to the Nazi movement, with evidence also suggesting that the Nazis were helped by extraterrestrials. David Icke's research suggests that the Iraq War of the 21st Century was influenced by the Reptilians.

Nuclear War

Throughout ancient writings, there are stories and legends about wars that could only be explained by the use of nuclear weapons. Only when one accepts the fact that advanced civilizations seeded our planet can we understand that these legends may have been speaking about nuclear weapons.

Zecharia Sitchin's research into the Sumerian writings discovered that the Anunnaki of pre-Flood times treated humans with little respect. The gods were jealous and often fought with each other. Humans were often used by the gods during their wars and were taught how to use primitive weapons, but the gods reserved their advanced weapons for themselves, weapons that suggest they were nuclear.

Around 2400 B.C., the ancient Indus or Harappan civilization was suddenly formed and collapsed just as suddenly around 1800 B.C. The writings of the Rig Veda describe the arrival of the Vedic Aryans to the Indus Valley in present day India. They were defeated in a battle by dark-skinned people living in walled cities. Scholars believe that most of the Vedas were in present form by 1500 B.C., but pre-1500 B.C. oral tradition had been going on centuries before. These ancient Vedic texts describe in detail the flying vehicle wars with frightful weapons used by the gods. One translation of an epic tale has Mahabharata describing several different species of extraterrestrials arriving in their celestial cars at a wedding feast. Included were the scaly Nagas, who were perhaps the Reptilians. Sri Lanka (Ceylon) was the stronghold of the Nagas, and additional writings describe Sri Lanka as being the home of strange Reptilian-like creatures.

The ancient Indian writings tell about flying machines called vimanda. The gods of Mahabharta were described as using missiles, antimissiles, and nuclear weapons that caused radiation sickness and sterility among the people of both sides of the terrible war. Sitchin gives evidence that the names of the evil gods of ancient Hindu literature were similar to the names of ancient Babylonian, Assyrian, and Egyptian gods.

Valdamar Valerian, an extraterrestrial researcher, writes in his book *Matrix II*, about a researcher named David Davenport who studied Hindu texts in depth. Davenport came to the conclusion that Mohenjo-Dara, which was the largest city of the Indus civilization, was destroyed by a nuclear blast. He found evidence in an agreement between the Aryans and space aliens to destroy the city, which was a city inhabited by the enemies of the Aryans, so the Aryans could continue mining minerals in peace. Before the city was destroyed, over 30,000 of its residents were warned to leave the city within seven days to escape the upcoming destruction.

As discussed in the Anunnaki chapter, Sitchin provides evidence that the two Biblical cities of Sodom and Gomorrah were destroyed by a nuclear blast. This resulted from a severe conflict between the first-born son of Enlil, named Ninurta, and Enki's first-born son, Marduk. Another nuclear blast occurred near the same time period at TIL.MUN on the Sinai Peninsula, which destroyed the spaceport. The radioactive fallout blew east to the city of Ur in Sumer, resulting in the decimation of the Third Dynasty of Ur, and many Sumerians suffered a horrible death. Following the nuclear blasts, the Anunnaki departed the area in great haste to avoid the radiation.

Sitchin's research into the Greek gods also determined that they fought nuclear wars. The Greek gods were extraterrestrials who fought other extraterrestrials on Earth. Much of Greek mythology regarding the gods is based on Sumerian and Egyptian legends. Sitchin believes that Zeus, who was the chief god of the ancient Greek civilization, used nuclear weapons when he fought the Titans.

Confirming all these ancient legends suggesting nuclear war, the Pleiadians told Billy Meier that there have been many nuclear wars on Earth. Billy was told that early civilizations, even prior to Lemuria and Atlantis, developed to a point where Earth humans and extraterrestrials quarreled, resulting in a nuclear war that destroyed civilization. Meier was told that a nuclear war depopulated North America about 50,000 years ago. They also confirmed nuclear blasts in Sodom and Gomorrah and in India, Lebanon, Australia, and Ecuador. The Pleiadians told Billy about "The Takauti Document of Japan," which predates all other records, going back 24,000 years. The teachings of the Shinto religion are based on this document, which also describes a nuclear war during this time period and even showed the location of each atomic blast and which cities were destroyed, using the symbol of a mushroom cloud.

The extraterrestrials have been both a blessing and curse to the planet Earth. They have advanced our standard of living through advanced technology on one hand, but have been responsible for death and destruction on the other hand. Some extraterrestrial civilizations seem not to be any more advanced spiritually than humans. Some of the more advanced extraterrestrial civilizations have suggested that they may prevent a nuclear holocaust in this era in order to allow Earth a chance to evolve into the next dimension. However, negative extraterrestrials want us to experience conflict and are working with the Secret Government to control and manipulate humanity to that end. Humanity still has free choice, and it is our responsibility to find the truth so that we can evolve and escape the confines of the Third Dimension and negative extraterrestrials. Many of the galactic civilizations are helping us in subtle ways so that we can make this giant leap in evolution.

PART II

THE EVOLUTION OF HUMANITY

Chapter Thirteen

THE SCIENCE BEHIND EVOLUTION
Getting the Hell out of Here

Earth humans are Three-Dimensional beings, and for decades, spiritual teachers have proclaimed how humanity may be evolving to the Fourth and Fifth Dimensions. While researching this book, the same message was coming to us from the extraterrestrials. We are approaching a window of opportunity that will allow us to enter these next two dimensions. It has to do with the 26,556-year cycle of the Precession of the Equinoxes, which will complete its cycle on December 21, 2012, a date which the Mayan calendar claims will mark the beginning of a new era.

Many of those who are spiritually enlightened know the soul is trying to evolve in order to get back to its Source. The soul is trying to perfect itself by learning lessons through various incarnations. On Third Dimensional Earth, we are embedded in the physical, and our souls want to leave the entrapment of Third Dimensional reality. The soul and humanity need to evolve to a higher consciousness in order to leave this Third Dimension. All matter is frequency, and the vibration of Earth has been increasing as we approach the end of the 26,556-year cycle. In order to evolve to the Fourth and Fifth Dimension, the extraterrestrials claim about eight percent of humanity needs to achieve this higher consciousness.

Many of the extraterrestrials that are interacting with Earth have evolved into higher dimensions and are trying to assist us in subtle ways that do not violate the galactic code of noninterference. Once Earth reaches this level of vibration, we can join our galactic family. The ascendency window will be from 2012 until 2022 A.D. Humanity has had seven other

239

opportunities to evolve when the 26,556-year cycle came to completion, and all these opportunities were missed. If this opportunity is lost again, humanity will have to wait for completion of the next 26,556-year cycle. Several factions of extraterrestrial civilizations are trying to impede our evolution and keep us imprisoned in the Third Dimension, as discussed in the earlier chapters.

The extraterrestrial information from the *Voyagers* series by Anna Hayes* provides about the best scientific speculation on our evolvement. Our evolvement involves DNA and the blocks that have been placed on it by extraterrestrials. It involves parallel universes and a planet in a different harmonic universe called Tara. It involves time portals and dimension portals. All these factors and more play a role in the principles underlying the process of ascension.

DIMENSIONS AND HARMONIC UNIVERSES

In order to explain dimensions and harmonic universes, we need to look at quantum physics. Everything is vibration or frequency. Higher frequencies can intermingle with lower frequencies. We know there are various frequencies in radio waves and television waves, and by tuning into the correct frequency, we can listen to the radio or watch television. All these wavelengths of different frequencies intermingle in the same space.

According to the extraterrestrials (from the Ranthkian civilization, Anna Hayes'* source), the universe is composed of five harmonic universes. Each harmonic universe is composed of 15 dimensions. All of these dimensions exist in the same space, but operate separately, because of the variant particle pulsation rates of the atoms. These particle pulsation rates take place in the same space, and at the same time, remain invisible. When a planet evolves through dimensions, the rate of speed or particle pulsation progressively increases, while the density of matter particles progressively decreases, allowing the planet to evolve through the 15-Dimensional scale. A track of time for Harmonic Universe One in which Earth exists equals 26,556 years.

* Anna Hayes is now writing under the name of Ashayana Deane

Planets have the ability to move from one dimension to the next and from one time continuum to the next. Each does this by magnetically drawing into its morphogenetic field particles from the Unified Field of Energy for each dimension. As a planet pulls in the frequency pattern of the next higher dimension into its morphogentic field, it will move upward into the next dimension. A morphogenetic field is a field of energy that provides shape to matter. Each dimension band has a specific rate of particle pulsation and a specific angular particle spin rotation.

This brings us to the concept of time. Time exists as a unified field of particles that pulsate at various rhythms and spin at various angles of rotation. This is what creates the illusion of manifested space and time, giving the appearance of individuated identities. Moving one's consciousness through a segment of the unified field of particle substance gives the illusion of space and time. Time does not move, but consciousness moves through the unified field of the Time Matrix.

This is but a primer on how extraterrestrials can move through space and time. Quantum physics has come to the conclusion that there is no such thing as space and time. Our consciousness creates the illusion. Presently, Earth and its humanity are in the Third Dimension of Harmonic Universe One, and we are trying to get to the Fourth and Fifth Dimensions. How can we do this? First we need to look at the planet Tara which operates in Harmonic Universe Two.

TARA

To understand how and why humanity arrived on Earth, we need to understand the planet Tara before Earth was even formed. We go back to Tara's history after it had been located in the First Harmonic Universe, where it successfully evolved into the Fifth Dimension and Second Harmonic Universe. The Turaneusiam race evolved for about eight million years on Tara into two main races, the Alanians and the Lumians. Both cultures evolved on Tara carrying the 12-strand DNA genetic code. The Alanians were controlled by an elite group called the Templar Solar Initiates who wanted to control the

Lumians. Many Alanians became aware of the misuse of power by the Templar Solar Initiates and defected to the Lumians and interbred with them. Approximately 550 million years ago, the power generator crystals exploded deep underground because of the Templar Solar Initiates' misuse of power. Portions of Tara's planetary grids were blown apart and detached from the morphogenetic field. The planetary fragments were drawn into Tara's sun, vaporized, and pulled into a black hole at the sun's center. It re-emerged into a galaxy of lower dimension in Harmonic Universe One. Twelve pieces of Tara's fragments began to build up matter density and remanifested their form because they still contained their portion of Tara's morphogenetic field. After entering Harmonic Universe One, the fragments formed our solar system. The planet Malduk imploded to the asteroid belt, and another planet, Nibiru, formed a large elliptical orbit around the sun. Tara's fragments then became part of the morphogenetic unified field structure of Harmonic Universe One. Souls of Tara had become trapped in time with fragmented units of consciousness within the unified field of Harmonic Universe One. A rescue mission was formulated for the lost souls of Tara, now living on Earth.

Five etheric cloistered races were used in the rescue mission. The five cloistered races were Ur Antrians, Breanoua, Hibiru, Melchizedek, and Yunaseti. These etheric races populated the Earth for many years. They began the first stage of seeding humans about 25 million years ago. Various racial mixtures were created through which the lost souls of Tara were able to ascend.

The morphogenetic field of consciousness energetically took on the shape of a sphere called the Sphere of Amenti, which was part of Tara's planetary core. The Sphere of Amenti was placed within the Earth's core, establishing a "wormhole" or portal link between Earth's core in Dimension Two and Tara's core in Dimension Five. The Sphere of Amenti created a stable portal structure allowing an open transit between Earth and Tara. The Sphere of Amenti was linked into the space/time of Tara's past, which would then be reattached to Tara's future time cycle.

There were six portals from the Sphere of Amenti that linked Tara's past through various time periods on Earth, called the

Halls of Amenti. These portals created a new time track through which Earth could re-evolve into the Fifth Dimension frequency band and merge with Tara. Once the merger occurred, Earth would become Tara on Harmonic Universe Two, and Tara would become Gaia in the next Harmonic Universe Three. These six portals in the Sphere of Amenti allowed the cloistered races to incarnate in the dimensional branch and time fields, two through six. This would pull into their consciousness and body forms the fragmented consciousness of the lost souls from each of the dimensional bands.

The lost souls of Tara originally had 12 DNA strands, and as they evolved in the dimensions on the Earth plane, the 12 DNA strands would assemble. The body form would progressively transmute into a less dense (higher vibration) form as it ascended into Harmonic Universe Two on Tara. Once passing through the dimensional portals, the incarnate would appear on Tara in a future time coordinate and become free from Harmonic Universe One. As the lost souls from Tara returned, so did the lost portion of Tara's morphogenetic field. The Sphere of Amenti's portal was a warp in time that allowed evolution to take place rapidly, and the Sphere of Amenti contained the mechanics and blueprint of human evolution. The time portals regulate the process of human evolution and their return to the Source. These are the dimensional passageways one must pass through to ascend Earth and move out of the Time Matrix.

The Halls of Amenti (the dimensional portals) have been a closely guarded secret by the priests of Ur and Mu (early Taran religions), their descendants, and the extraterrestrials for millions of years. When the portals were created 25 million years ago, the priests on Tara drew out from Tara's Fifth Dimensional core a pattern of frequency that represented the morphogenetic field for the entire planetary grid. Once a Harmonic Universe One planet pulled its portion of the higher frequency, it could undergo dimensional ascension into Tara's grid, replacing its portion of Tara's morphogenetic field. Once each of the Harmonic Universe One planets of the solar system ascended back into the Tara grid, Tara could ascend into Harmonic Universe Three and become Gaia.

Earth's portion of the morphogenetic field was called the

Blue Flame or Staff of Amenti and it represented the key to evolution of Earth and human lineage. It is the gateway into Tara's morphogenetic field, and in Biblical terms it as called "The Pearly Gates of Heaven."

ASCENSION

When Earth merges with Tara, it will mark the coming of the Fifth World of Native American legend. Earth's ascension is a highly involved scientific process. Ascension is the path of order that consciousness evolves through a structural multidimensional system. It involves "multidimensional energy mechanics representing universal order through which consciousness experiences as being." In other words, one is conscious when moving from one dimension to another. The process involves the transmutation of particles and antiparticles into a progressively less dense rate of matter, where a planetary body is able to evolve from a lower dimension frequency band into a higher band within the 15-dimensional scale.

To understand ascension, one must understand morphogenetic fields, which is the form-holding construction that allows matter and antimatter to build into individual forms. To fully ascend through the 15-dimensional scale, the energy structure must assemble all portions of its original morphogenetic field. A morphogenetic field is a tapestry of interwoven energy particles composed of literal substance. The ascension process involves assembling morphogenetic fields into their original pattern that had been fragmented from Tara. In other words, it recombines the field into its original form of the energetic tapestry in the next dimension.

Particles and antiparticles are composed of units of multidimensional sounds or tones; within each frequency band there are base tones and overtones. Ascension involves the process of merging particles and antiparticles to create the merging of frequency patterns. This process brings together base tones and overtones that emerge out of the same morphogenetic field. As base tones and overtones merge, a resonant sound is created, allowing the merging of particles and antiparticles that transmute into pure energy. It is the process of building dimensional frequencies through merging of particles and antiparticles that allows matter to form and evolve up through the dimensional scale.

Through this ascension process Earth and Tara will merge. Following the merger, changes will take place within the body matter of Earth. Planetary ascension takes place in waves over millions of years. At the time the matter-particle planet passes through the grid of its antimatter double from the parallel universe, there is a fusion of energy that causes a morphogenetic wave to be sent out to the above dimensional band. This wave will project its particle and antiparticle form, manifestating within the new dimension and corresponding time cycle.

As a planet reaches the fusion point, the people of the planet have also reached the fusion point and will undergo particle transmutation. However, only a portion of the human population will be able to transmute and ride the planetary wave of ascension. Only those individuals who have fully assembled the Third Dimensional and Fourth Dimensional frequency will ascend on the wave into Tara's energy tapestry. The decision to ascend will be decided by the soul and not by the personality. Only the greater soul identity knows when the consciousness has reached a high enough level to properly ascend. Also, a natural soul-orchestrated death will allow the consciousness to ascend if the physical person has died and the greater soul deems it time for ascension. The morphogenetic field of these people will merge with Tara's Fourth, Fifth, and Sixth Dimensional particles and antiparticle structure. As a result, the matter particles will re-manifest into the Second Harmonic Universe. They have now entered into a future version of Earth with a new time cycle. This ascension wave will take place between 2012 - 2022 A.D.

MECHANICS OF ASCENSION

Ascension is the scientific process of the evolution of consciousness and biology that is governed by the laws of energy mechanics when applied to a multidimensional reality system. After death, our consciousness lives on and evolution continues. Eventually, all souls will ascend through the Fifth Dimensional scale and re-emerge as a sentient identity within the realm of pure consciousness that is beyond the dimensional systems.

Humanity was created as a creator species. This means that the thoughts and actions entertained by humans are manifested on Earth and in the life experience beyond the Earth's plane. The choices one makes in thought and deed determine one's feeling toward the quality of that particular experience.

Our evolutionary blueprint, called the "Covenant of Palaidor," is stored within the morphogenetic field of the Sphere of Amenti, where the organizational plan and purpose is also kept. The blueprint has guided us forward toward an unseen destiny, which is immortality and the reunion of consciousness with our creative Source. Immortality and freedom from death, diseases, and pain are the natural birthright for the human species. The life challenges we all face are lessons in growth as human consciousness evolves to remember its eternal existence. The choices we make can bring us joy, harmony, good health, and freedom. The Sphere of Amenti (the morphogenetic field of consciousness) is important in our evolution as a species, because it holds the blueprint through which our race consciousness manifests. The Sphere of Amenti connects the human species to the greater morphogenetic field of the Earth, which is connected to the morphogenetic field of planet Tara in Harmonic Universe Two. In order to evolve out of Harmonic Universe One into Harmonic Universe Two, the frequency Dimensions of Four, Five, and Six from Harmonic Universe Two must be manifested in the energetic grid of Earth. The sound tones must become operational within the active DNA strands of humans. The frequencies contained within that imprint will determine the placement of consciousness following death. Humans have the power to evolve consciously and to develop the bioenergetic system to build the needed frequency into the DNA. It is our personal responsibility.

According to extraterrestrial sources, DNA is the key to humanity's evolution. DNA is the abbreviation for deoxyribonucleic acid found in genes that are located in the nucleus of cells. It is the molecular basis of heredity. Chemically, DNA is a series of nucleic acids constructed in the form of a double helix strand held together by hydrogen bonds between purine and pyrimidine bases. Scientists have been able to discover that much of our DNA has no function, so they have labeled it junk DNA. This junk DNA may actually be the key to our evolution. Extraterrestrial sources have emphasized that in order to evolve, we must activate more strands of DNA. How is that possible if we only have two strands of DNA? According to extraterrestrial information, there are latent strands of DNA found in our morphogenetic field that can be activated. This all relates back to a sister planet in Harmonic Universe Two, called Tara, from which the Earth originated. In the next chapter, we shall discover the extraterrestrial role in seeding the planet Earth from their 12-strand DNA.

The greater the amount of the 12-strand DNA imprint that can be activated within the body's DNA, the more conscious awareness and multidimensional knowledge will be available to the physically embodied consciousness. To end the cycle of rebirth in Harmonic Universe One, an individual must attain a Fifth Dimensional frequency. Individuals who die without assembling the Fifth DNA strand imprint will continue evolution in the Fourth Dimensional astral planes. Those individuals who have not fully assembled lower dimensional strands will have to go through experiences in Harmonic Universe One to assemble the lower strand.

Individuals born into bodies with advanced genetic lines had to evolve through incarnations carrying the smaller gene code imprint in order to attain the frequency of consciousness compatible with the larger gene code body. Those with the larger gene code have earned the privilege of evolutionary progress. Each individual is at a different level of evolution. As humans with larger gene code packages assemble their Fifth DNA stand, they pull higher dimensional frequency into the Earth's grid, assisting the Sphere of Amenti to open into the Earth's core morphogenetic field. The Earth, in turn, can transmute Fifth Dimensional frequency into the bioenergetic field of everyone. If the Earth's vibration is too low, those with higher frequencies cannot birth into the planet. Therefore, the vibrational frequency of Earth determines the evolution of DNA in humans. A lower Earth vibration cannot hold the Sphere of Amenti. In order to ascend, the Earth must sustain a concentration of souls with a higher frequency gene code in order to hold the Sphere of Amenti or morphogenetic field of consciousness.

THE TIME CYCLES OF EARTH

During the 26,556-year cycle of time, there are only certain points in time that the Earth's grid system reaches a vibrational rate high enough to receive an infusion of Fifth Dimensional energy. Only at these times can Earth sustain a larger concentration of high frequency souls. During these periods, the Sphere of Amenti can fully open to allow assembly of the Fifth DNA strand for all the races. It can only occur four times during the 26,556-year cycle. During this window, the energy grid of Earth can fuse with the energy grid of its double within the parallel universe as will its counterpart parallel double Tara from the next harmonic universe. The two 4,426-year cycles at the beginning and end of the 26,556-year cycle are called an "Ascension Cycle." During the second Ascension Cycle that we are now going through, the Earth's subatomic particles that are composed of overtone frequencies of Dimension One, Two, and Three will merge with the antiparticles of Tara of the base frequencies Four, Five, and Six within the parallel dimension scale of Harmonic Universe Two. When the Earth's particles fuse with Tara's antiparticles, both the particle and antiparticle enter hyperspace. At that moment they return to their higher dimensional morphogenetic fields. This releases a wave of particle/antiparticle energy called a morphogenetic wave. The merger of Earth's particle and Tara's antiparticle in hyperspace transmutes them into their respective dimensional morphogenetic field. The antiparticles of Tara become magnetic and the particles of Earth become electrical. The two planetary grids merge on Earth as etheric overtone structures. A new land configuration will manifest physically on the particles of Earth, and the land of Tara will rise up on Earth over a progression taking 2,213 years. At the height of the wave crest, the magnetic field temporarily collapses, and polarity will be reversed, causing a naturally occurring time warp and portal passage that will emerge between Earth and Tara.

Each harmonic universe, including Harmonic Universe One, has a harmonic time cycle of 26,556 years. Each cycle is synchronized through the subparticle pulsation rate within each of the 15 dimensions. A time warp portal opens up all the way through the 15-dimensional scale. The interdimensional portal opening creates a morphogenetic wave that allows a beam of UHF (ultra high frequency) light particles to pass through the harmonic universes. This is called a Holographic Beam, which feeds and sustains all of the 15 dimensions. The portal opens during the half point of the Ascension Cycle and again at the close of the Ascension Cycle 2,213 years later. Because Earth is connected to the Pleaidian star system, the Holographic Beam follows a path through the Pleiades to Earth. Alcyone is the central sun of the Pleaidian star system to which the Holographic Beam is connected. The photon energy around the Pleiades is referred to as the photon belt, which is residual energy left over from the last morphogenetic wave. This energy is replenished every 22,256 years as the Holographic Beam projects through Alcyone two times during the second Ascending Cycle.

LEFT BEHIND

During the five-year period prior to 2017 and after (2012-2017), the imprint of the Fifth DNA strand will be made available to all races in order for ascension to Tara. The DNA has to be fully activated in order to ascend. During this ten-year window beginning in 2012, the Halls of Amenti portal passage to Tara becomes open to the masses. The Sphere of Amenti must be held open in Dimension Two of the Earth's morphogenetic field in order for the morphogenetic wave, mass ascension, and evolutionary leap in time to occur. In order for this to happen, eight percent of the population must embody the Earth's morphogenetic field (Blue Flame) for ascension to occur.

Those whose body energy and consciousness pulsate at the Fifth Dimensional frequency will perceive a direct, body transmutation through a time portal from Earth to Tara, or they may receive time travel transport to Tara by way of an interdimensional space-craft. The latter group will include those who cannot make it through dimensional transmutation of the portals, and they will end up in Tara's space/time coordinate about 5,232 years in the future from Earth's space time position. This is beginning to sound like the rapture of Christian mythology.

Those who cannot fully assemble the Fifth DNA strand will be left behind, as the Halls of Amenti will no longer be passable. These individuals will be stuck in the Harmonic Universe One incarnation cycle until the Fifth DNA strand can be built, or until the next morphogenetic wave occurs 2,213 years later. The ten-year period, 2012 - 2022, allows humanity a great opportunity for an evolutionary leap in time.

There will be many human souls who cannot assemble the Fifth DNA strand, but who have assembled the Fourth DNA strand. Many will choose to drop their physical bodies so they can reassemble Dimension Five frequency into morphogenetic fields from Dimension Four. These individuals will ascend to Tara as soul essences from Dimension Four. Those humans who have the highest vibration rates will reenter hyperspace and reappear on Earth within Harmonic Universe Two, 5,532 years in the future. During the morphogenetic wave 2,213 years in the future, which is the close of the Ascension Cycle, the remaining base tone material of Earth will transmute, and Earth will shift into Dimension Four time scale.

For the Halls of Amenti to open for ascension during the ten-year window, the Earth grid vibration must be raised high enough to open the Arc of the Covenant, an event that will be discussed more fully in the next chapter. The Arc of the Covenant is a portal bridge between the Sphere of Amenti and the Andromeda galaxy. The Arc of the Covenant was designed so that the Sphere of Amenti could eventually be reentered to access the Earth through the portal bridge of the Arc. During the third seeding of humanity, the Sphere of Amenti had to be removed from the Earth's core because of diabolical extraterrestrials, but the Arc of the Covenant provided access through a portal bridge.

If the Blue Flame, Earth's morphogenetic field, is not embodied on Earth when the grids on Earth and Tara begin to intersect, the infusion of Dimension Five energy from Tara and the Holographic Beam cannot be run through the Earth's grid during the crest of the morphogenetic field. As a result, the Earth and humanity will be trapped for another 26,556 years within Harmonic Universe One. Humanity would regress to Dimension One, Earth's lowest vibration, and would again have to evolve for another 26,556 years. This digression and repeating of Harmonic Universe One time has already occurred seven times in Earth's evolution.

Chapter Fourteen

THE SEEDING OF EARTH
A Remarkable History

This chapter will show how our galactic family planted the seeds of humanity on Earth and nurtured its development. At times, too many weeds crept in, resulting in the destruction of civilization. Being good gardeners, our galactic forefathers seeded Earth again, only to have the weeds return. Not being disheartened, they seeded the garden a third time, and, hopefully, this time humanity has learned its lesson from the past to allow the harvest of souls into the Fourth and Fifth Dimensions. This metaphor of seeding refers to the introduction of humans on Earth, and we are presently in the third seeding. The path has been difficult for humanity, but we again have a chance to grow to maturity for a fruitful harvest.

The history of ancient Earth is fascinating, and as we shall see, history tends to repeat itself, even if there have been millions of years between seedings. The Native Americans talk of five worlds. The First World represented the original Turaneusiam culture of Tara, the lost souls that entered Earth many millions of years ago and had to be rescued by our galactic family. To begin the rescue, etheric cloistered races populated the earth for many ages, creating various racial mixtures through which many of the lost souls of Tara were able to ascend. This era was called the Second World. The next two worlds involved the physical seeding of Earth that carried the genetic material from many star systems. Humanity is an eclectic society of races, thanks to our galactic ancestors. The material in this chapter is largely based on Anna Hayes'* *Voyager Series*, that was given to her from an extraterrestrial source. It is the most detailed history of ancient

252

Earth that I have seen, and, hopefully, I can make this complex material understandable to you, the reader. Earlier in the book, we looked at evidence that supports this remarkable knowledge. Let's return to Tara to help sort out our ancient history.

THE TARA CONNECTION

We discussed in the previous chapter how a cataclysm occurred on Tara in Harmonic Universe Two. About 550 million years ago, the planet Tara fragmented into 12 large pieces that were sucked into a black hole and reemerged in Harmonic Universe One. The morphogenetic field (the energy field that provides shape to matter) of these fragments formed our present-day solar system. Along with the Taran fragments was the matrix of souls that found themselves trapped in Harmonic Universe One.

A rescue mission was devised to rescue these Taran souls so they could continue their evolutiont. The plan became known as the Covenant of Palaidor formed by the Sirian Council, and the Elohim, Ceres, Lumians, and Alanians. Collectively this group was known as the Palaidorians.

Several members of this group have not been discussed. The Elohim, created by the Lyrans in Harmonic Universe Three, became the overseers of the Sirian races in Harmonic Universe Two. The Elohim had interbred with the Cerrasz Turaneusiam subrace on Tara, creating a race called the Ceres. A powerful priesthood arose out of Ceres called the Priesthood of Mu. They were a spiritual collective with a strong matriarchal doctrine centered on the sacred Law of One, or unity consciousness. The original Taran race, who had 12 strands of DNA that was later enhanced, was called Turaneusiam. This revitalized race became known as the Ur-Tarrantes, who created the Priesthood of Ur upon the continent of Mu on Tara. Both the Priesthoods of Mu and Ur exist today and serve as gatekeepers of the time portal between Earth and present-day Tara. The Lumarians and Alanians were two civilizations on Tara whose counterparts on Earth later became Lemuria and Atlantis.

* Anna Hayes is now writing under the name of Ashayana Deane

The First Seeding

The plan to rescue the Taran souls formulated by the Palaidor civilizations involved five smaller morphogenetic spheres that originated from the morphogenetic field of the Sphere of Amenti (the morphogenetic field of consciousness). These spheres became the morphogenetic patterns for the five races known as the cloistered races, also known as the Palaidorians who represent the Earthly counterparts of the larger Palaidorian group on Tara.

These five morphogenetic fields became five physical Root Races Three, Four, Five, Six, and Seven that evolved from these five etheric cloister races. The five cloistered races seeded the brown (Ur-Antrian), the red (Breanoua), the white (Hibiru), the yellow (Melchizdek), and the black (Yunaseti) races. The plan called for each Root Race to evolve and assemble one DNA strand, strands two through six. Their etheric cloister race would hold the imprint for strands seven through twelve. On Earth, the physical races would follow their etheric counterparts, the cloister races.

The human body, which was originally immortal, was designed to transmute and ascend, not to die and reincarnate. Originally there were 24 cycles of transmutation, which later became 24 reincarnation cycles of death and rebirth. As the consciousness passed from one life to the next, the DNA would assemble, so the soul could learn its lessons on its path of evolution.

In the first wave of physical creation on Earth, there were 60 humans that were manifested out of the cloistered races' morphogenetic forms. All five cloistered races were represented in equal standing, and the Earth's population progressed through this lineage. The cloistered races populated the Earth for many generations, creating a variety of racial mixtures through which many of the lost souls of Tara could ascend.

The Third Root Race and first physical race to appear on earth were the Lumarians, a brown race that appeared 15 million years ago. They assembled the second strand of DNA. Following the Lumarians came the Fourth Root Race, the Alanians who appeared about 9 million years ago. They were a red-skinned race who were responsible for assembling the Third DNA strand. The Alanians created the subconscious and Dimension Two emotional body.

The five cloister races and Lumarians and Alanians evolved together on Earth between 2.5 million and 5.5 million years ago. Those who kept the integrity of the genetic code through each of the races were able to ascend within their immortal bodies through the Halls of Amenti portals. According to native American legend, this era represented the Third World, which was the first seeding.

About 5.5 million years ago, many of the races began to interbreed with animals. As a result, they lost their ability to transmute genetically and thus lost their immortality. To complicate things even more, a great war broke out involving higher universe entities, called the Electric Wars, which lasted 900 years. The resultant treaty closed the Halls of Amenti portals between Earth and Tara because of the contaminated genetic codes of Root races Three and Four. The sealed portals meant humanity had to evolve into the morphogenetic field of their race and pick up DNA strands seven through twelve. Without the imprint frequency patterns of strands four through six, the incarnate soul could not plug strands seven through twelve into the operational genetic code. The portal of the Halls of Amenti had now been sealed (called the Seal of Palaidor) to human lineage by removing the morphogenetic field of the Third and Fourth cloister races. Another problem had been created, as there was an energetic block between the physical and etheric bodies within the bio-energetic auric field. This resulted in a perception of duality that then existed between consciousness and body.

In summary, as a result of the ramifications from the Electric Wars and interbreeding with animals, the Elohim created a Fourth Dimensional frequency block within the Second and Third DNA strand. This sealed the Third and Fourth Root Races (Lumarians and Alanians) out of their morphogenetic field/soul matrix, creating a build-up of soul fragments in Dimensions Two through Four that would have to be integrated into Root Race Five consciousness. It was this seal that created the subconscious mind. Root Race Five now needed to integrate the soul fragments of the Lamanians and Atlanians of the second seeding and the Lemurians and Atlanteans of the third seeding to be discussed. Things were not good on Earth following the Electric Wars. They were so bad, in fact, that Earth could no longer sustain life.

Earth had become an ascension planet to higher dimensions, so souls were able to achieve dimensional ascension through re-evolution. They did this by re-evolving back into their original 12-strand DNA body type, thanks to the Covenant of Palaidor and the Sphere of Amenti. The Seal of Palaidor, which sealed the ascension portals, created a problem within the incarnational process and for the physical incarnates on Earth. Because of the seal, the race memory of those souls incarnating was wiped out of the planet's cellular memory. Humans were unable to remember where they came from, their life purpose, or where they were going on their evolutionary path. Essentially, the races of the second seeding entered an incarnation with no memory of their identity. This knowledge was stored in the Fourth Dimension and could be accessed only through astral essences, such as intuition. Humans of the second seeding had a new type of consciousness, a perception of exaggerated duality and a sense of separation that included forgetting the truth of the Law of One, or unity consciousness. The Seal of Palaidor put Earth in a frequency quarantine, disconnecting it from the galactic family. Because of this exaggerated duality, the Sphere of Amenti was also removed from the Earth's core about 5.5 million years ago.

The Second Seeding

Approximately 4 million years ago, the Sirian Council, the Elohim, and other Harmonic Universe Two groups developed a plan to restart human evolution. The plan was to allow souls in Dimension Four to reenter Earth by a process called downgrading. The vibrational essence of the race morphogenetic field could be stepped down to a slower vibrational pattern by passing a portion of the morphogenetic field through a core of another Harmonic Universe One planet. The planet chosen was Sirius B. This resulted in a hybrid strain of Sirian/human consciousness that developed in souls to be known as Kantarians. Later, the Kantarians founded the Kantarian Federation that served as a guardian and education race for the humans of the second and third seedings. They became heavily involved with the old Sumerian and Egyptian cultures as well as Atlania in the second seeding and Atlantis in the third seeding.

During the second seeding era, the portal bridges between Sirius B, the Sphere of Amenti, and Earth allowed for greater options for the Fourth and Fifth Races, which carried within their morphogenetic field the third DNA strand. Those soul essences who had assembled most of the third DNA strand could move into the morphogenetic field of Sirius. About 4 million years ago, the race morphogenetic field for the second seeding races was entered into Sirius B.

The races began incarnating on Earth through a small group of human hybrids. This group had found exile within the Pleiades Star System and evolved there during the Electric Wars. They then migrated to Sirius B where they interbred with the Kantarian race. A smaller hybrid race was formed called the Dogos, which strengthened the human and Sirian genetic code. During the second seeding, the Dagos, Kanatarians, and the Pleiadian lineages became intertwined with the original Turneusiam (Tara) imprint.

Root Race Three (Lumarians) of the first seeding became the Lamanians of the second seeding, denoting passage through Sirius B. Root Race Four (Alanians) became known as the Atlanians in the second seeding, 2.5 million years ago, and Atlantis in the third seeding. The Fifth Root Race Aerians and their cloister Hibiru entered the second seeding about 1.5 million years ago. The era beginning with the second seeding and extending through the third seeding represented the Native Americans' Fourth World. The Fifth Root Race Aeirians (later known as Ayrians) were responsible for assembling the Fourth DNA strand, corresponding to Fourth Dimensional frequency. This energy field allowed the Sphere of Amenti to be returned to the Earth's core. As they fulfilled the assemblage of the Fourth DNA strand, the Seal of Palaidor would be lifted, allowing soul fragments from Dimensions Three and Four to merge with Fifth Race consciousness and continue their evolution. All the second seed races carried the genetic code of Pleiadians and Sirius Dogos.

Within the operational DNA strands of the second and third seedings were grounding codes. These codes contained partial frequency patterns for each of the 12 dimensions within the 12-strand Taran imprint. The grounding codes allowed for transmutation of physical matter particles, once the overtone activation codes became operational. Once DNA strands One through Four plugged into each other, the matter particles would begin to fuse with their antiparticles. This would activate the Fifth, Sixth, and Seventh grounding codes and allow the body to transmute through the portals of the Halls of Amenti to reappear on Tara.

Approximately one million years ago, things on Earth took a turn for the worse when a race of extraterrestrials arrived from the Harmonic Universe One Orion star system. They tampered with the genetic codes of the races and altered the evolutionary imprint for many Amenti souls. Hybrids were created within the root races called Dracos. These were intelligent, sentient, and upright-standing beings. Appearance wise, they looked like lizards on their hind feet with a human body and aggressive, warlike behavior. They exhibited poor relations with the Harmonic Universe One Galactic Council and had little respect for any life forms, showing a lack of spiritual development. In reality, the Dracos' lineage was a mutated hybrid that evolved in the Orion star system. They felt they were entitled to use the planet Earth as they wanted because they were members of human lineage. The Dracons also carried the 666 genetic configuration of the Templar Axion Grid, denoting certain DNA and overtone blocks.

Originally, extraterrestrials had seeded dinosaurs on Earth 375 million years ago. The Dracon geneticists tampered with dinosaurs and bred carnivorous and aggressive species. Dracons and their creations became a menace to humanity. A plan by some guardian extraterrestrials and humans was concocted to eliminate this problem. This occurred about 956,500 years ago, when the Anunnaki extraterrestrials from Sirius A and human races tried to destroy the underground habitats of the Dracons, but the plan backfired, creating a pole shift. Following this catastrophe, most of the Dracons left or were destroyed by the ensuing climatic changes. However, they left a legacy of genetic distortion that threatened the human lineage. Because of Dracon contamination, humanity began to degenerate further as interbreeding with animals became a serious problem.

Atlanians of the second seeding were influenced by descendants of the Sirian-Anunnaki race from Harmonic Universe Two, which later evolved for a time on Harmonic Universe One, Mars. This was the same race that had originally caused the great cataclysm on Tara. The Anunnaki had brought to Atlania the creed of the Templar Solar Initiates from Tara, which was an elitist, sexist, and materialistic distortion of the Law of One. Women were viewed as subservient to men and were used as breeders of hybrid children. About 950,000 years ago, the Anunnaki created a race called the Nephilim, which included the genes of humans, Atlanians, and Anunnaki.

A great war broke out on Earth involving the Anunnaki, with a resultant treaty allowing the Anunnaki to advance spiritually through the teachings of the Law of One. Not all of the Anunnaki agreed to the treaty, and this rebel group known as the Anunnaki Resistance continued to be a threat on Earth. The Dracos hybrids formed an alliance with the Anunnaki Resistance as did their Dracon forefathers. They conspired to destroy the Sphere of Amenti. After hearing about the plan, the Sirian Council took steps to control the actions of the Anunnaki and Dracos hybrids by administering the Templar Axion Seal to their morphogenetic field. The seal prevented interbreeding with other species who did not carry the imprint of their gene pool.

About 900,000 years ago, the Sphere of Amenti was returned to the Earth's core under the direction of the Elohim, Sirian Council, and Harmonic Universe Two Palaidorians. This accelerated the assemblage of the DNA strands and released the DNA Seal of Palaidor from some races. In addition, several groups had evolved enough to release the DNA Seal of Amenti, allowing ascension through the Halls of Amenti.

About 849,000 years ago, the Anunnaki had developed a plan to destroy the Sphere of Amenti in order to use humans as a working race. After learning about their plan, the Guardians removed the Sphere of Amenti once again from Earth to another location. As a result of this drama, the Sirian Council formed the Sirian-Arcturian Coalition to protect the Sphere of Amenti, Earth, and its allies. This organization was invited into a larger group called the Interdimensional Association of Free Worlds, whose members included those from high harmonic universes. Once the Sphere of Amenti was removed from Earth, great climate and geographical changes took place, including a great flood. Most of humanity was destroyed. When all these cataclysmic Earth changes settled down, there was yet to be a third seeding.

Third Seeding Ascension

Because the Sphere of Amenti had been removed to the Andromeda galaxy from the Earth's core, a portal bridge was constructed between Earth and the Sphere of Amenti. It was called the Arc of the Covenant, which allowed the race souls of the Sphere of Amenti to be reseeded. The guardians of Earth had placed a Fifth Dimensional Seal upon the Arc of the Covenant that allowed souls to only descend to Earth, not ascension to the Sphere of Amenti morphological field. To ascend, they would require a Fifth Dimensional coding in their gene structure. Evolution would take time for Root Races Three, Four, and Five of the third seeding. Souls who could not reenter the Amenti morphological field would be able to evolve within the Fourth Dimensional astral plane between incarnations. The Fifth Dimensional Seal on the Arc of the Covenant was designed to be released if eight percent of the population assembled the Fifth DNA strand. Then Earth's grid vibration would rise, which would open the Earth's morphogenetic field to accept the Sphere of Amenti. At that time, everyone's DNA would be stimulated and the lost memory of multidimensional reality would reenter the cellular memory of individuals. The Blue Flame of Amenti would descend to Earth within the Sphere of Amenti, again opening up an ascension portal to Tara.

Third Seeding Civilizations

Many individuals of the second seeding who had found refuge underground following the cataclysmic events on Earth prospered within Inner Earth's communities. Other groups managed to survive within the tunnel system under the Earth's surface created by their ancestors. Several groups of the now extraterrestrial humans returned to Earth and began preparing for the third seeding, which was to be orchestrated through the Arc of the Covenant portal bridge. Two primary groups of returning humans orchestrated the breeding, descendants of the Seres Egyptians and the Hebrew (Melchizedeks/Hibiru hybrids). The Third Race, the Ur-Antrian etheric clositer, began birthing about 75,000 years ago, followed by the physical Lamania Root Race 2,000 years later. They were seeded on a land mass, Lemuria, in the Pacific Ocean. The Breanoua clositer of the Fourth Root Race entered 72,000 years ago followed by the physical Atlanians 70,000 years ago, who seeded a continent in the Atlantic Ocean, which became known as Atlantis. About 68,000 years ago, the Fifth Race cloister Hibiru entered the host matrix followed by the Root Race Aryan 3,000 years later. They were seeded into the region of today's Black Sea and Carpathian Mountains. Another seeding was accomplished by the Annu-Melchizedek into the Atlantean race with the Egyptian subrace through the Melchizedek cloister matrix about 68,000 years ago.

The group of Anunnaki who had joined the Sirian Council, along with the Sirian-Katarians and Sirian Blue Race of Sirius B, were the three Sirian races that greatly influenced the Atlantean culture. On the other side of the planet, the Pleiadian races of Harmonic Universe One were assisting the civilization of Lemuria.

The cultures of the third seeding thrived under the influence of the advanced extraterrestrial races. The Annu-Melchizedek organized their civilization around the spiritual and scientific teachings of the Law of One. As the Earth's civilization prospered, the Sirian Council allowed the Atlantean and Lemurian civilizations to receive technology from the advanced extraterrestrial races. Large crystal power generators were given to Atlantis and Lemuria as a gift from the Sirian Blues of Sirius B to allow energy to be drawn directly from the Earth's core. The Blue Flame of Amenti allowed the Atlantean civilization to access multidimensional frequencies through which the morphogenetic field of Earth's matter could be directly affected. As a result, Earth's gravitational pull could be neutralized, so objects could be manifested and demanifested, and even teleported to any desired location.

The Annu and Hebrew people of the Melchizedek cloisters Host Matrix had become the primary guardians of the Arc of the Covenant. Until about 55,000 years ago, Earth civilization thrived by following the Law of One. Things were going well for the planet until members of the Anunnaki Resistance appeared and infiltrated the Atlantean culture. They created genetic and social digression that nearly ended the Atlantean civilization.

The Destruction of Lemuria

During the height of the Lemurian and Atlantean civilizations, a conspiracy to overtake the planet was developed by the Anunnaki Resistance and their Dracon allies. They wanted to overthrow the planet when civilization had reached its maturity. The Anunnaki Resistance had interbred with the Annu-Melchizedek on Atlantis and distorted the genetic line. The sacred teachings of the Law of One were slowly distorted and replaced by the creed of the Templar Solar Initiate of Tara, creating division and social unrest in society.

During this time period, the Dracos had secretly returned, infiltrated Lemuria and created an extensive network of underground tunnels and lairs. From these, the Dracos began terrorizing the Lemurians as well as the Atlanteans. The Lemurians hoped to use the power crystal generators to create a small underground explosion to rid Lemuria of the Dracos. The plan backfired and a massive explosion occurred beneath the landmass of Lemuria, causing the continent to rip apart. Many of the Lemurians had retreated underground after the explosion and some permanently entered the Inner Earth. Most, however, returned to the Earth's surface to rebuild. Atlantis then became the cultural center of the planet, but it never reached the peak Lemuria had during its prime. Most of the Dracos were evacuated by their Anunnaki Resistance accomplices.

Atlantis

Following the cataclysm, the Sirian Council and their allies returned to Earth to help humanity rebuild. When it was discovered that the portal passage to the Arc of the Covenant had been damaged in the explosion and the energy grid beneath had become unstable, the Elohim, the Ra Confederacy, and other Earth guardian groups were able to restructure the Arc of the Covenant and move the portal to Egypt from Atlantis. From Egypt, the Arc was able to recharge the Atlantis crystal generator with multidimensional frequencies.

At this time, conditions continued to get worse on Atlantis. The Templar Annu used their influence to corrupt the Atlantean priests, which led to a decay of morals and the social structure. Another plan was concocted by the Anunnaki Resistance about 48,500 years ago. This time they were going to use the generator crystals to break through the electromagnetic barrier protecting Inner Earth. After discovering the plan, the Sirian Council instructed the Sirian Anunnaki to make a great show of power in Egypt and Atlantis against the Anunnaki Resistance by protecting the portal opening to the Inner Earth. Through the portal, the Arc of the Covenant could be accessed. To provide protection to the portals, the Sphinx and pyramids were built, with pyramids being the hallmark of Anunnaki architecture and scientific achievement.

The Sphinx was the first monument created, covering the portal to the Inner Earth that was linked to the Arc of the Covenant. The Sphinx had the body of a lion and the head of an Anunnaki warrior and served as fortification for the Inner Earth portal. It also housed a great energy transmitter that was charged with UHF (ultra high frequency) Fifth Dimensional energy from the Arc of the Covenant. Historically, this machine later became known as the Ark of the Covenant. The Anunnaki also had a small tool in the shape of an ankh, allowing high frequency energy to be synthesized in specific ways. Ankhs were used to create the Sphinx and pyramids by providing power to reverse gravitational pull and to directly affect the particle makeup of the morphogenetic field of Earth matter substance, helping to explain how the pyramids were built.

The next major structure built by the Sirian-Anunnaki was a monumental pyramid located directly over the portal opening of the Arc of the Covenant. The Great Pyramid was originally built about 48,500 years ago, rebuilt in 10,500 B.C., and again in 9,000 B.C. The design and construction involved the Anunnaki of the

Sirian Council, Annu-Melchizedeks, Hebrews, and Seres-Egyptian people. Besides fortifying the portal of the Arc of the Covenant, it also served as an interdimensional teleportation center.

The Great Pyramid was originally designed as a harmonic resonance center through which a multidimensional frequency band could be pulled in from deep space, especially from the planetary core of Siris B. Both the Great Pyramid of Giza and the Arc of the Covenant were located on the energy vortex, which happened to be on the geographical center of Earth, the heart chakra of Earth. This vortex allowed an energy exchange between Earth and Fourth Dimensional frequency bands. The creation of the Great Pyramid allowed immediate intervention by the Sirian Council and Galactic Federation if the Arc of the Covenant should come under attack by the Anunnaki Resistance. From 46,459 B.C. to 28,000 B.C., the Great Pyramid was an active interstellar teleport station. For those who had the correct DNA, the Great Pyramid was also used for ascension, if approved by the Elohim.

Since their creation, the Sphere of Amenti (550 million years ago) and the Arc of the Covenant (840,000 years ago) were the primary focus of human evolution. The process and purpose of the human evolutionary imprint was held within the Sphere of Amenti. These means of soul evolution needed to be protected from the Templar-Anu.

As the Templar-Anu were formulating their plan to conquer Inner Earth with the crystal generator, they were becoming more hostile, as many had relocated to Atlantis. Around 28,000 B.C. they carried out their plot. They forced power through one of the many generator crystals, when the crystal unit exploded, creating more than ten times the energy of an atomic bomb. Because the Earth's energy grid had been weakened from the Lemurian explosion, this latest explosion tore the continent of Atlantis apart and reduced the land mass to residual islands.

A number of Atlanteans had been forewarned of what might be happening and had relocated to Egypt. Many went underground and became part of Inner Earth communities. As a result of the explosion, the Earth tilted on its axis, causing more planetary destruction.

The aftermath was devastating to the planet, besides the great earth changes. Because of the tilt, the Great Pyramid could no longer be used as an intersteller teleport station. The Sirian Council and Anunnaki could not respond as rapidly to help the planet because they could not use the Great Pyramid as an interstellar teleport station. Following the cataclysm, the people began to regroup on Atlantis, and the Law of One began to reemerge between 28,000 B.C. and 12,500 B.C. Angered by their loss of power in Atlantis, the Anunnaki Resistance launched an aggressive assault against the Sirian Council outpost on Mars and then in Egypt. Their intention was to destroy the Arc of the Covenant and take control of Earth's civilization. Following the destruction of the Mars outpost and the destruction of the Sphinx and Great Pyramid, the Sirian Council and their allies were able to put down the Anunnaki Resistance upheaval. They were driven from Earth to the distant planet Nibiru that was orbiting the sun in an elliptical orbit.

Both the Great Pyramid and the Sphinx were rebuilt. The Great Pyramid was now aligned with the Pleiades star system and the planet Alcyone, which allowed it to once again be an interstellar teleport station. The Sirian Council decreed that the Anunnaki Resistance and the Dracos were not permitted access to the Alcyone spiral for Earth visitations. However, unauthorized groups could use the intergalactic portal system that connected to the Earth's portal system.

The remaining Atlantis civilization degenerated even more into a tortuous, elitist group run by the Templar-Anu, who were controlled by the Anunnaki Resistance. Because of their actions about 11,500 years ago, Earth civilization was changed forever. The Templar-Anu devised a plan to destroy the Sphere of Amenti. Without race morphogenetic fields, the souls on Earth would be trapped in Harmonic Universe One, and their evolutionary imprint would be erased.

The Templar-Anu wanted to create a worker race that would supply them with natural resources, especially gold. To achieve their goal, they were going to experiment genetically with humans. To destroy the Sphere of Amenti, they planned to direct UHF Five Dimensional energy from an ankh through the great crystal generator. In 9558 B.C., they miscalculated, and the first explosion ripped apart the largest Atlantean island and triggered a tectonic plate shift below the ocean floor, resulting in the sinking of all the islands of Atlantis.

The Sirian Council had no choice but to order the Sirian Blue Race to remove all the unexploded generators on Earth. Upon removal, humanity's technology was thrust into the dark ages and civilization became very primitive. The ankh technology was banned from the surface and taken to Inner Earth, and the technology would not be returned to Earth until humanity matured.

A serious problem arose on Earth after the great explosion. The electromagnetic pulse from the great crystal generator had sparked open the original seal on the Arc of the Covenant, which kept the Sphere of Amenti in place. From Andromeda, the Sphere of Amenti began its descent to Earth that took 2,000 years, arriving in 7558 B.C. and missing the dimensional blend of 9048 B.C. and 6835 B.C. If the Sphere of Amenti tried to enter the Fourth Dimension vortex on Earth, the sphere would not open but would explode upon impact with the natural frequency barrier separating the Third and Fourth Dimensions. The impact would cause the Fourth Dimensional vortex chakra to collapse, and the buildup of energy would cause the Earth's grid to explode.

The Sirian Council and the Elohim devised a plan to prevent this catastrophe. They created a natural frequency fence or energy barrier to surround the Earth and prevent the Sphere of Amenti from passing. The plan was to keep the fence in the astral plane until the Third Dimensional frequency reentered the Earth's morphological field.

One disadvantage of the frequency fence was that the natural structure or interstellar portals, which allowed interstellar transit to Earth, would be cut off from the interdimensional grid. It would be difficult for visiting space-craft to find an easy way to Earth. Another decision made by the Council was to let humans orchestrate their own affairs, with the guardian extraterrestrials overseeing the evolutionary ascension behind the scenes.

Essentially, Earth and humanity were put under galactic

quarantine. They lost all direct assistance and relationship with multidimensional intergalactic communities. Information from higher dimensions would no longer transit into Earth's grid and cellular memory. Over the millennia, the Earth's involvement with galactic civilizations was lost, as was the high degree of technology they had once known.

Another dilemma resulted from the frequency fence. The soul essence passing through the Earth's core for birthing could not ground the full imprint for the Third DNA strand. The Third Dimensional tones would not allow the companion codes from the soul's morphogenetic field to plug into the Earth's morphogenetic field. The frequency fence caused the Third DNA strand to manifest without certain base tones and overtones, resulting in a genetic mutation for the human lineage. As a result, a new kind of consciousness developed within the races.

The portion of personal identity that manifested through the lower base tones of the Third DNA strand became the lower self, or ego. The higher octaves of the Third DNA strand manifested that part of the identity called the Higher Self. This resulted in a conscious mind that was divided into two parts that did not consciously associate with each other. The ego, or lower mind, became obsessed with the five sensory perceptions and developed an exaggerated sense of dualism. The ego was responsible for aggression and had a need to dominate. Because of the frequency fence, the ego had been cut off consciously from its personal morphogenetic field and the morphogenetic field of its race and planet. As a result, human consciousness was locked into the illusion of matter and could not comprehend the reality of nonmanifested substance from which all manifested substances are created. In other words, consciousness lost touch with its Source.

To heal the egotistical mind would require integration with the Higher Self. Over time, the Higher Self allowed the Fourth and Fifth DNA strand to begin manifesting. It could communicate only to the consciousness by way of subconsciously sensed feelings or intuition.Today, the Third DNA strand is beginning to reverse mutate through the lifting of the frequency fence. The Higher Self will merge with the ego, once the Third DNA strand is completely assembled. Then mental awareness will comprehend itself as a multidimensional identity. The process will be accelerated if the ego allows the Higher Self to speak. Once the higher and lower minds become integrated, a person will have developed a superconscious mind.

This chapter has shown how history repeats itself throughout the three seedings. We are approaching the end of a great cycle and have the opportunity to evolve to a higher dimension.

We do this by trying to understand the Law of One, that we are all connected. To evolve, we need to practice forgiveness, have compassion, be of service, and come from a place of love, not fear. Let us not waste this rare opportunity for all of humanity's sake.

REFERENCES

Andrews, George, *Extra-Terrestrials, Friends and Foes*, Illuminet Press, Lilburn, GA, 1993.
Chapters 5, 8, 9, 10, 11, 14

Bernard, Raymond, *The Hollow Earth*, Citadel Press, Secaucus, NJ, 1969.
Chapter 9

Boulay, R.A., *Flying Serpents and Dragons: The Story of Mankind's Reptilian Past*, The Book Tree, Escondido, CA, 1997.
Chapter 4

Bramley, William, *The Gods of Eden*, Avon Books, New York, 1989.
Chapters 1, 11, 12

Coleman, John, *Conspirators' Hierarchy: The Story of the Committee of 300*, American West Publishers, Carson City, NV, 1992.
Chapters 10, 11

Collier, Alex, Internet
Chapters 8, 10

Corso, Philip, *The Day After Roswell*, Pocket Books, New York, 1997.
Chapter 12

Essene, Virginia and Sheldon Nidle, *You Are Becoming a Galactic Human*,
S.E.E. Publishing Company, Santa Clara, CA, 1994.
Chapters 3, 8

Hayes, Anna, *Voyagers: The Sleeping Abductees, Volume I*,
Granite Publishing, Columbus, NC, 1999.
(Anna Hayes is now writing under the name of Ashayana Deane)
Chapters 1, 4, 5, 8, 9

Hayes, Anna, *Voyagers: The Secret of Amenti, Volume II,*
Granite Publishing, Columbus, NC, 1999.
(Anna Hayes is now writing under the name of Ashayana Deane)
Chapters 4, 13, 14

Horn, Author, *Humanity's Extraterrestrial Origins: ET Influence on
Humankind: Biological and Cultural Evolution,*
Silbur Shnur, Lake Montezuma, AZ, 1996.
Chapters 5, 10

Hurtak, J.J., *The Keys of Enoch,* The Academy for Future Science,
Los Gatos, CA, 1977.
Chapter 12

Icke, David, *Alice in Wonderland and the World Trade Center Disaster,*
Bridge of Love Publications, Wildwood, MO, 2002.
Chapters 4, 11

Marrs, Jim, *Alien Agenda,* Harper Collins Publishers, New York, 1997.
Chapters 10, 11

Marrs, Jim, *Rule By Secrecy,* Harper Collins Publisher, 2000.
Chapters 10, 11

Milanovich, Norma, *We, the Arcturians: A True Experience,* Athena
Publishing, 11827 E Cannon Dr., Scottsdale, AZ 85259
Chapter 6

Nichols, Preston B., *Experiments in Time,* Sky Books, New York, 1992.
Chapter 10

Pinkham, Mark, *The Return of the Serpents of Wisdom,*
Adventures Unlimited Press, Kempton, IL, 1997.
Chapter 4

Robbins, Diane, *Messages From the Hollow Earth*,
    TGS Services, Frankston, Texas, 2003
    Chapter 9

Roberts, Diane, *Telos*, Mount Shasta Publishing, CA, 1996.
    Chapter 9

Royal, Lyssa and Keith Priest, *The Prism of Lyra*, Research Press,
    Scottsdale, AZ, 1990.
    Chapter 1

Royal, Lyssa and Keith Priest, *Visitors From Within*,
    Research Press, Scottsdale, AZ, 1992.
    Chapter 5

Salla, Michael, *Exopolitics: Political Implications of the Extraterrestrial
    Presence*, Dandelion Books
    Chapter 8

Sitchin, Zecharia, *The Twelfth Planet*, Avon Books, New York, 1976.
    Chapter 7

Sitchin, Zecharia, *The Stairway to Heaven*, Avon Books, New York, 1980.
    Chapter 7

Sitchin, Zecharia, *Wars of God and Men*, Avon Books, New York, 1985.
    Chapter 7

Starr, Jelaila, *We are the Nibruians: Return of the 12th Planet*,
    Granite Publishing, Columbus, NC, 1999.
    Chapters 4, 7

Stevens, Wendelle and Stefan Denaerde, *UFO Contact from Planet Iarga*,
    Stevens Publishing, Tuscon, AZ, 1982.
    Chapter 4

Temple, Robert, *The Sirius Mystery: New Scientific Evidence of Alien Contact 5,000 Years Ago,*
Destiny Books, Rochester, VT, 1998.
Chapter 3

Walden, James, *The Ultimate Alien Agenda; The Re-engineering of Humankind,* Llewellyn Publications, St. Paul, MN, 1998.
Chapter 4

Winters, Randolph, *The Pleiadian Mission: A Time of Awareness,*
The Pleiadian Project, Yorba Linda, CA, 1994.
Chapters 2, 8

INDEX

INSTITUTE FOR THE STUDY OF GALACTIC CIVILIZATIONS

1304 South College Avenue
Fort Collins, Colorado 80524
www.galacticcivilizations.org

Mission Statement: To educate the public about galactic civilizations.

Statement of Purpose: The purpose of the Institute for the Study of Galactic Civilizations (ISGC) is to educate the public about Earth's galactic visitors, their culture, and their influence on humanity. Such influences include the areas of spirituality, religion, science, politics, government, war, and health. The Institute will also raise awareness about the galactic message regarding a raising of human consciousness and a potential dimension shift. Education will be accomplished through workshops, conferences, lectures, website, newsletters, and press releases. The Institute will also provide research and support to individuals who have experienced galactic contact and will cooperate fully with other organizations who are exploring galactic civilizations.

Financial Support: The Institute for the Study of Galactic Civilizations is a nonprofit organization (501-c-3) that relies on financial donations.

Information: For information about conferences, workshops, galactic news items, and galactic contacts, please contact the Institute at *www.galacticcivilizations.org.*

GALACTIC CONTACT

If you have had contact with our Galactic Visitors and want to share your experience, please contact the **Institute for the Study of Galactic Civilizations**. The Institute is creating an archives of galactic interactions. Also, if you would like emotional support regarding your experience, the Institute will supply you with a list of counselors.

Institute for the Study of Galactic Civilizations

1304 South College Avenue
Fort Collins, Colorado 80524
www.galacticcivilizations.org

OTHER BOOKS BY ROBERT SIBLERUD

Sacred Science Chronicles

Volume I   *In the Beginning: Mysteries of
Ancient Civilizations*

Volume II   *Keepers of the Secrets: Unveiling the
Mystical Societies*

Volume III *The Science of the Soul: Explaining the
Spiritual Universe*

Volume IV   *The Unknown Life of Jesus: Correcting
the Church Myth*

New Science Chronicles

Volume I   *Our Future Is Hydrogen: Energy,
Environment, and Economy*

Volume II   *Mind Your Health, Heal Your Body: A Guide
to Wellness*

New Science Publications
9435 Olsen Court
Wellington, CO 80549

IN THE BEGINNING

MYSTERIES OF ANCIENT CIVILIZATIONS

The Sacred Science Chronicles Volume I

Robert Siblerud

Foreword by: Maury Albertson, Ph.D.
(Former Director of Research at Colorado State University)

In the Beginning offers a new perspective on the role ancient civilizations played in shaping human society over the millennia. It begins with the most ancient of civilizations, Lemuria, the lost continent of the Pacific, and explains why the ancients called it the "mother of civilizations." Lemuria seeds spread to the civilizations of Atlantis, Egypt, the Mediterranean, and the Americas. This unique book gives evidence that the Celts, Egyptians, Libyans, and Phoenicians inhabited North America long before Columbus.

The gods of these ancient civilizations played an important role in their development. *In the Beginning* explains who the gods were and where they came from, suggesting their extraterrestrial origins. Sacred science helps explain the many mysteries of these ancient civilizations, giving a new perspective to *Genesis*.

In the Beginning is Volume I of the *Sacred Science Chronicles*. Sacred science is the science that examines all universal laws, laws that go beyond the physical. It operates on the axiom that a spiritual dimension exists that functions under the laws of the universe.

KEEPERS OF THE SECRETS

UNVEILING THE MYSTICAL SOCIETIES

The Sacred Science Chronicles Volume II

Robert Siblerud

Foreword by:" Brian O'Leary, Ph.D.
(Former Apollo astronaut)

Keepers of the Secrets gives the reader a unique perspective of mystical societies throughout civilization. It summarizes the history and spiritual philosophy of the shamans, Druids, Essences, Gnostics, Hermetics, Kabbalists, alchemists, magicians, witches, Sufis, Rosicrucians, and Freemasons. The book provides insight on how these societies have influenced the world through the centuries.

These mystical societies shared many spiritual truths that contradicted orthodox beliefs. As a result, the church and state tried to suppress most of these societies, forcing them to become secret as humanity was not ready for many of these truths.

THE SCIENCE OF THE SOUL

EXPLAINING THE SPIRITUAL UNIVERSE

The Sacred Science Chronicles Volume III

Robert Siblerud

Foreword by: Leo Sprinkle, Ph.D.
(Professor Emeritus University of Wyoming)

The Science of the Soul describes the nature and purpose of the soul. Followers of most religions do not understand the principles underlying the soul. Age-old questions of life's purpose, why we are here, and human destiny are answered by understanding the soul. Scientific evidence helps prove the existence of a spiritual world, and in these pages, the reader will journey through these spiritual realms.

By examining various religious beliefs about the soul, *The Science of the Soul* takes the reader into depths that underlie these religious doctrines. The book explains the involution (the Fall) and evolution of the soul, the concept of rebirth and destiny, the spiritual planes including Heaven and Hell, and the spiritual inhabitants and their hierarchy. The treatise also examines the soul after death, spirit communication including electronic techniques, and finally the holographic science underlying spirit and matter.

THE UNKNOWN LIFE OF JESUS

CORRECTING THE CHURCH MYTH

The Sacred Science Chronicles Volume III

Robert Siblerud

Foreword by: Maury Albertson, Ph.D.
(Co-founder of the Peace Corps)

The Unknown Life of Jesus shares the secrets about Jesus that have been kept from people for two millennia. The time has come to release this knowledge to people who seek spiritual truths and wish to escape Church dogma. Readers will discover how the early Church created the myth about Jesus by incorporating pagan beliefs and doctrines from other religions. It was perpetuated through corrupt editing of the New Testament. Early manuscripts written in Jesus' language of Aramaic have been discovered and contradict many of the writings of the New Testament.

Included in the book are childhood stories of Jesus currently unknown to most people. The lost years of Jesus are also explained, giving evidence that Jesus studied the teachings of Buddha and Hinduism while in India and was married to Mary Magdalene who bore Him children. Evidence also shows that Jesus survived the crucifixion and continued His teaching in Kashmir, where He is buried today after living well past 100 years.

Over the centuries, keepers of these secrets were the Priory of Sion in France, who claim to be the founders of the Knights Templar, precursors to the Freemasons. Today, the Priory is releasing this hidden knowledge to the world, including the genetic legacy of Jesus.

OUR FUTURE IS HYDROGEN!

ENERGY, ENVIRONMENT, AND ECONOMY

The New Science Chronicles Volume I

Robert Siblerud

Foreword by: T. Nejat Verziroglu, Ph.D.
(President of International Association for Hydrogen Energy)

Our Future Is Hydrogen! Is a book that gives our planet Earth hope by providing a solution to global warming and dwindling fossil fuels. Its purpose is to provide public awareness to this remarkable energy carrier. Both the automobile and oil industries have realized that hydrogen will be the energy of the future, beliefs held by many politicians and world governments. These industries will be major players in our hydrogen economy.

Describing the planet's need for an alternative energy source, *Our Future is Hydrogen!* shows why hydrogen is the best of the alternatives. The reader will appreciate hydrogen's value by learning of its properties, history, safety, applications, and current status. The transition to a hydrogen economy has just begun, accelerated by California's dedication to a clean environment.

MIND YOUR HEALTH, HEAL YOUR BODY

A GUIDE TO WELLNESS

The New Science Chronicles Volume II

Robert Siblerud

Mind Your Health, Heal Your Body explores the reasons why people become ill, the mind/body connection, and alternative methods for healing. The modern day health care system has a difficult time treating the cause of many illnesses, most often treating only symptoms. Because Western health care operates under a misguided paradigm regarding health, it has become one of the largest causes of illness and death in the United States, largely due to excessive use of drugs, surgery, and unnecessary diagnostic tests. The book in contrast, looks at Eastern medicine, examining its holistic approaching of treating mind, body, spirit, and emotions, and the proper flow of qi energy. Western science is beginning to understand the principles underlying Eastern medicine.

Optimum health depends upon a good immune system that can be adversely affected by the mind, emotions, and stress. For example, the book gives clear evidence that most back pain is caused by repressed emotions. If our mind can cause illness, the mind can also heal the body, as demonstrated by the placebo effect, biofeedback, and hypnosis. Western medicine ignores the spiritual relationship to health that the book explores, suggesting that there may be a spiritual meaning underlying many illnesses.

NEXT BOOK

WE ARE ONE

THE SCIENCE OF UNITY CONSCIOSNESS

The Sacred Science Chronicles Volume VI